SpringerBriefs in Computer Science

More information about this series at http://www.springer.com/series/10028

Seng W. Loke

Crowd-Powered Mobile Computing and Smart Things

 Springer

Seng W. Loke
School of Information Technology
Deakin University
Burwood, VIC, Australia

ISSN 2191-5768 ISSN 2191-5776 (electronic)
SpringerBriefs in Computer Science
ISBN 978-3-319-54435-9 ISBN 978-3-319-54436-6 (eBook)
DOI 10.1007/978-3-319-54436-6

Library of Congress Control Number: 2017932755

Printed on acid-free paper

This Springer imprint is published by Springer Nature
The registered company is Springer International Publishing AG
The registered company address is: Gewerbestrasse 11, 6330 Cham, Switzerland

Preface

This book came about from my own exploration (together with collaborators) of a range of topics in mobile and pervasive computing. Particular themes tend to occur often in mobile and pervasive computing research, including context-awareness, cooperative systems, crowd computing, smart things and cloud computing, calling for a book that tries to identify key concepts and synergise ideas in one place, under *crowd-powered mobile computing and smart things*. Much of the work discussed in this book draws from my own published work (with collaborators) and references are given where applicable—there are also many sentences in the book prefixed by 'might', 'can be' and 'could be' which tend to state unexplored ideas or food for thought, rather than prescribing solutions. If current technological and end-user trends are to continue, there are only going to be more smart things, smarter things, more people with such things, larger crowds with such things as the densities of living spaces increase, and greater interconnections among them, yielding an unprecedented Web of synergistic links and innovation possibilities.

The book is intended for researchers, students or anyone interested in the topics outlined in the first chapter, from mobile cloud computing, Internet of Things, drones to swarm cooperative systems. The book does not aim to be a deep treatise on these topics, but aims to be an overview, and to connect ideas, so that some topics are only touched on. There are still many relevant topics including data science, human-computer interaction, and cybersecurity implications, not discussed, but the idea of a brief is to be brief, so that some topics are inevitably left out.

I would like to acknowledge the many collaborators, including researchers and doctoral students, with whom a lot of the work discussed in this book has been explored, and to them and to readers: 'may there be more voyages of discovery!'. At the end of the day, a book is only one perspective on the area.

Deakin University   Seng W. Loke
Melbourne, Australia
January 2017

Acknowledgements

The author would like to thank La Trobe University and Deakin University, where the author worked while this book was being written, for providing an excellent environment for research and study.

> *Where no counsel is, the people fall: but in the multitude of counsellors there is safety.*
> – Proverbs 11:14 (KJV)

> *Without counsel purposes are disappointed: but in the multitude of counsellors they are established.*
> – Proverbs 15:22 (KJV)

Contents

1 Ubiquitous Connections: The Internet of People and Things 1
  1.1 Introduction 1
  1.2 Technology Trends: An Overview 1
    1.2.1 Cloud Computing 1
    1.2.2 Internet of Things (IoT) and the Device Mesh 2
    1.2.3 Big Data 3
    1.2.4 Big Crowds with Increasingly Powerful Mobile
        and Wearable Computers 3
    1.2.5 Crowd Computing, Crowdsourcing (or Human
        Computation) 4
    1.2.6 Culture of Sharing 6
    1.2.7 Collective Computing 7
    1.2.8 Swarm Dynamics 8
  1.3 Five Ideas and This Book 8
  References 9

2 Crowd+Cloud Machines 11
  2.1 Combining Crowd and Cloud Computing 11
  2.2 Types of Mobile Clouds 12
  2.3 Characteristics of Crowd+Cloud Machines:
    The Case of Honeybee and Multi-Layered Honeybee 13
  2.4 Decentralised Spatial Computing with the Crowd 15
  2.5 Spatial Finding with the Crowd 18
  2.6 CAROMM and GroupSense: Crowdsensing and Crowd
    Activity Recognition 19
  2.7 Crowd+Cloud Machines to Assist People with Disabilities 21
  2.8 Physical Annotation Systems.. 21
  2.9 Summary 22
  References 22

3   Extreme Cooperation with Smart Things 27
   3.1   Things Cooperating More Than Ever 27
   3.2   Vehicle-to-Vehicle Cooperation 28
      3.2.1   Benefits of Cooperation to Reduce Traffic Congestion 28
      3.2.2   Cooperating over Time 29
      3.2.3   Cooperation to Resolve Contention for Car Park Spaces 32
   3.3   Interactions and Relationships in Cooperative Living
      Room IoT: Device Ecologies 33
   3.4   Cooperation Within Large Crowds: The Case of Crowd Steering 35
   3.5   Cooperation for Sharing Things: Decentralised? 35
   3.6   Correlated Equilibrium 36
   3.7   Summary 37
   References 38

4   Scalable Context-Awareness 39
   4.1   Context-Aware Mobile Computing..................................... 39
   4.2   Larger Scale Sensing: Place Level Sensing 40
   4.3   Social Sensing... 40
   4.4   Scaling Up Context-Awareness.. 41
   4.5   Scalable Context-Awareness for Smart Cars: A Use Case 43
   4.6   Summary 45
   References 45

5   Drone Services for Mobile Crowds 47
   5.1   The Rise of Drones... 47
   5.2   Can We Imagine Drone Services?...................................... 47
   5.3   Issues and Challenges.. 49
      5.3.1   Scheduling Drones.. 49
      5.3.2   Sharing and Shifting Control of Drones 50
      5.3.3   Smart Things Interacting with Drones 50
      5.3.4   Infrastructure to Support Drone Services....................... 51
      5.3.5   Drones from the Crowd 52
   5.4   Summary 53
   References 54

6   Social Links for Crowds and Things 55
   6.1   Favour Networks in Mobile Crowds 55
   6.2   Automatic Social Networking 59
   6.3   Social Networks for Things 61
   6.4   Summary 63
   References 63

7   Conclusion and Future Work 65

Chapter 1
Ubiquitous Connections: The Internet of People and Things

1.1 Introduction

Urban populations around the world are increasing and cities are becoming increasingly complex. This creates challenges for how people live and work, given limited resources. However, while the crowds increase in size and density, they are also more connected via mobile devices than ever before. Billions of increasingly sophisticated networked mobile devices provide an increasingly significant resource, surrounding each user as well as spanning communities. Such a resource is giving rise to new opportunities in mobile computing, impacting and integrating its associated areas including context-aware computing, crowdsourcing, mobile services, mobile cloud computing, social networks and the Internet of Things (IoT). The aim of this book is to provide a perspective on how mobile computing will take shape, as powered by the crowd for the crowd.

1.2 Technology Trends: An Overview

There have been a number of technology trends that are shaping computing today. Building on over 20 years of mobile and ubiquitous computing research, as well as technological developments in a range of related areas, a non-exhaustive list is as follows.

1.2.1 Cloud Computing

Large companies such as Amazon, Google and Microsoft are able to achieve economies of scale and sustained reliability in providing compute resources, storage and a range of infrastructure services at competitive prices, giving rise to commodity clouds. But services and offerings have been rising in levels of abstraction providing functionality that takes users away from the need to manage servers and handle low level details—the often called *serverless computing* notion, where developers need not care or worry about managing the servers and their underlying operational details.[1] Amazon Web Services (AWS) Lambda[2] and Google Cloud Functions[3] enable developers to focus on creating functions to handle events without worrying about capacity scaling and managing resources.

At the same time, there have been developments in what can be called the *'new cloud'* that is situated between the end-users (with mobile devices) and the remote cloud servers (e.g., building-sized data/server centres). This 'new cloud', which relates to mobile clouds [8, 9], cloudlets [22], edge-clouds [3], and fog computing,[4] provides resources with potentially reduced latency and greater security (as data is transferred and often kept only within a local network) by providing scalable resources closer to end-user devices, especially when the large remote cloud resources are not required. Also, nearby mobile and IoT devices, with their increasing capabilities, can form device clouds, or clouds of things, in order to provide resources to each other or to other devices.

Future abstractions could hide the actual locations and operational details of resources as long as seamless operation and elasticity in resource provisioning can be achieved, towards a broader view of the 'cloud' in cloud computing.

1.2.2   Internet of Things (IoT) and the Device Mesh

IoT has been an emerging topic of interest in industry and academia. A vision of 500 billion things connected to each other or to the Internet by 2025, attributed to John Chambers, a former CEO of Cisco, seizes the imagination of many. Embedding computation, networking and sensing into familiar everyday objects (from shoes to the umbrella) is only part of the story. New things will emerge in the future [11, 14, 21],[5] different from anything we previously knew, will form part of the IoT network, possibly connected and organised in various clouds of clouds (and so on) of things. The Internet of Flying Things has also captured the imagination of many, with drone technology maturing, and more drone services and applications emerging.

---

[1] https://developer.ibm.com/openwhisk/what-is-serverless-computing/,   http://searchitoperations.techtarget.com/definition/serverless-computing.

[2] https://aws.amazon.com/lambda/details/.

[3] https://cloud.google.com/functions/.

[4] http://fognetworks.org.

[5] http://ttt.media.mit.edu.

Gartner names *device mesh* as one of the top ten technology trends for 2016,[6] where a device mesh

> refers to an expanding set of endpoints people use to access applications and information or interact with people, social communities, governments and businesses. The device mesh includes mobile devices, wearable, consumer and home electronic devices, automotive devices and environmental devices - such as sensors in the Internet of Things (IoT).....As the device mesh evolves, we expect connection models to expand and greater cooperative interaction between devices to emerge.

The device mesh emphasises the connectivity among devices and things, as well as the potential for cooperation among such devices in order to perform a range of tasks, once they are networked to each other.

1.2.3 Big Data

With the proliferation of things with sensors and connectivity, things will, in effect, generate data at a high rate as they sense yielding a proliferation of data streams that either needs to be analysed in-situ or sent to the cloud or a cloudlet for processing—the need to manage, store only what is required, and make sense of the data as efficiently and safely as possible is an interesting challenge [17].

1.2.4 Big Crowds with Increasingly Powerful Mobile and Wearable Computers

There are more active mobile devices than people on earth since late 2014,[7] and some countries like Australia has more mobile phone accounts than its population since early 2015.[8] With tablets and smartphones and other devices coming on board, there will large crowds of people with mobile phones. Mobile devices are deemed just as important and widespread in some developing and third world countries, e.g., it has been noted that mobile phones are more important than food and electricity in Africa,[9] and mobile phones can be crucial to survivor for refugees.[10] However, new types of devices are increasing, not just mobile phones, and all these devices are riding on Moore's law, that is, the processing power and memory (RAM) increase

---

[6]http://www.gartner.com/newsroom/id/3143521.

[7]http://www.cnet.com/news/there-are-now-more-gadgets-on-earth-than-people/.

[8]http://www.sbs.com.au/news/article/2015/01/02/australia-has-more-phones-people.

[9]http://www.outwardon.com/article/cell-phones-in-africa/.

[10]http://theconversation.com/phones-crucial-to-survival-for-refugees-on-the-perilous-route-to-europe-59428.

significantly every year,[11] and the number of cores per device is also increasing (up to 16 is on the horizon at the time of writing of this book). While such devices will be multifunction and increasingly, applications are being designed to make use of idle time and resources, there is effectively a large supercomputer comprising the devices in the pockets of the audience in a large concert hall, mostly idle, and similarly, at night, there is potentially a large supercomputer comprising all the devices in an apartment building, also largely idle. Such computational resources could be exploited or harnessed if a seamless secure software infrastructure could facilitate this (of course with energy budgets and resource constraints imposed on alien jobs ran on devices, and proper compensation to device owners).

Mobile computing is broader than smartphones, smartwatches and tablets. There are also drones, smart cars and smart buses, and all sorts of new wearable things and everyday objects such as smart toothbrushes, smart hair brushes[12] and smart forks[13] that are computers which are literally mobile.

1.2.5   Crowd Computing, Crowdsourcing (or Human Computation)

The term crowdsourcing was coined by Jeff Howe, a contributing editor at Wired magazine [13], defining it as the 'act of a company or institution taking a function once performed by employees and outsourcing it to an undefined (and generally large) network of people in the form of an open call.' Since then, the notion of human computation or crowdsourcing has been substantially developed with the popular example of reCAPTCHA and many other ecosystems of crowd workers in complex workflows [18]. Crowdsourcing applied to data management is already an expanding area of research [15, 16].

But it is not only crowds of people with their smartphones that can provide both human computation and machine computation, but also cars. Consider the cars parked at an airport long term carpark for days, or cars parked at a shopping centre for long hours, there is potentially a supercomputer there if their computational and storage resources can be safely pooled together (with bounds on energy consumption so that their battery power remain adequate). Cars within an area can also cooperate (e.g., via vehicle-to-vehicle communication) and exchange information forming a multinode sensor network throughout an urban area, for example—even parked cars can participate as sensors forming a distributed (mobile) sensor network with nodes (i.e., the vehicles) sharing information, and thereby, achieving situation awareness beyond what each individual car can achieve, however sophisticated it is.

---

[11]http://www.mooreslaw.org.

[12]E.g., https://www.withings.com/us/en/products/hair-coach.

[13]E.g., https://www.hapi.com/product/hapifork.

1.2.5.1 Mobile Crowdsourcing

Smartphones equipped with mobile devices function as probes into the world, a mobile distributed sensor network situated with their owners, yielding the notion of crowdsensing [10], and mobile crowdsourcing [7, 12, 19, 20, 27]. In mobile crowdsourcing, users can respond to micro-tasks by doing them manually (e.g., translate a chunk of text or perform a task such as looking up something by going there physically), and can also contribute information such as their mood and other data to help create a spatiotemporal map of the world. Someone sitting at a stadium watching a soccer match might want pictures taken of the game from multiple perspectives throughout the stadium—instead of walking around to take photos, s/he could become an aggregator and crowdsource for photos: like-minded people seated in different positions throughout the stadium will be the 'workers', taking their own pictures just where they are, but also sharing them, each 'worker' could also be an aggregator him/her-self and so, multiple people get a tapestry of images made up by photos (and videos perhaps) from people throughout the stadium.

1.2.5.2 Spatial Crowdsourcing

Related to mobile crowdsourcing is the idea of *spatial crowdsourcing*,[14] where people need to move to particular locations to perform a task. gMission[15] is such a platform which features a collection of techniques including geographic sensing, worker detection, and task recommendation to get information related to geographic locations. The work in [25, 26] on spatial crowdsourcing considers server matching tasks to workers taking into account travel costs between workers and the location where tasks are to be done. Spatial crowdsourcing with workers selecting tasks are considered in other work, e.g., [6]. The trajectories of people's movements are also context which can be exploited for efficient crowdsourcing of urban logistics tasks, with workers tolerating limited deviations from their paths [5].

1.2.5.3 Social Machines

The combination of technology and people in an organised framework can be generalised into what has been called *social machines*[16]:

> Social Machines are a characterization of technology-enabled social systems, seen as computational entities governed by both computational and social processes.

---

[14]http://research.microsoft.com/en-us/projects/spatialcrowdsourcing/.
[15]http://www.gmissionhkust.com/.
[16]http://sociam.org/social-machines.

Social machines include social networks, citizen science social machines like Galaxy Zoo, knowledge sharing social machines like Wikipedia and public service social machines such as Ushahidi and Crime reports.[17]

1.2.5.4   Participatory Systems

Similar to the notion of social machines but emphasising the centrality of participation for the system to work is what has been called *participatory systems*, enabled by people with connectivity, primarily via their mobile devices. A definition from TU Delft is as follows[18]:

> Participatory systems are large-scale social-technical systems enabled by technology/connectivity, coordinating and orchestrating self-organisation, designed to provide individuals and organisations the ability to act and take responsibility in today's networked society.

This relates to crowdsourcing—e.g., urbanites provide feedback on their cities. From governance for smart cities to volunteered geographic knowledge [23] and citizen science, the notion that the citizens (or urbanites, in the case of a city) can co-create, together with administrators and government, their environment and their cities, or contribute to science, is an attractive idea.

1.2.6   Culture of Sharing

From bike sharing, car sharing (e.g., made possible by systems like Uber[19] and Lyft[20]), home sharing (e.g., Airbnb[21]), meal sharing,[22] to umbrella sharing,[23] as well as social sharing (e.g., via social networks such as Facebook and others), there is an interesting emerging culture of sharing and putting one's own resources, which have largely been in the domain of private usage, into public use, facilitated by suitable incentives and a platform [4] (e.g., payment, reputation, accountability, advertisement, and search). As the Internet of Things develop, *Thing sharing* might become more commonplace, with private things and places, and excess resources, being turned into services via a suitable platform. This notion of sharing is becoming a new way to organise economic activities [24], decentralising somewhat from large

---

[17]https://www.crimereports.com.

[18]http://www.participatorysystems.nl.

[19]http://www.uber.com.

[20]http://www.lyft.com.

[21]https://www.airbnb.com.

[22]https://www.eatwith.com.

[23]http://umbrellahere.com.

corporations, towards *crowd-based capitalism*.[24] It remains, however, to be seen what the limitations of sharing would be, e.g., whether sharing smaller and cheaper objects or more private items will be more trouble to manage than is beneficial. Such sharing is made more interesting and useful only when a sizeable crowd of people participate.

Incentives for sharing are diverse and can range from monetary benefits to social esteem or simply altruism. A car that is privately owned but shared (e.g., its Uberisation or via carpooling) so that its utility is increased can contribute towards society in the sense that it can reduce the need for others to get a car, and could make the overall transport system more efficient if low occupancy vehicles on the road can be reduced. A car can also offer other resources for rent, from its compute power to its external (e.g., digitised) surface (for advertising). A range of questions remain about such sharing, or thing sharing in general. If (autonomous) cars can be easily harnessed in multiple ways to make money, would it increase or decrease car sales? How much will private car ownership lead to a tragedy of the commons, as individuals reap greater benefits while shared transport infrastructure gets stressed, and will sharing alleviate or exacerbate the issue? Optimists would advocate sharing and, going further, even the commoditization of cars (or computer/communication devices) where mobility services (or, correspondingly, communication/computation services) take centre stage—however, demand for differentiated services is not likely to go away.

It seems that there is a trend towards what can be called 'we'-centric computing as opposed to 'me'-centric computing, where, instead of computing centred on the individual, one buys a device and becomes part of a giant collective, and at the same time, participates in a large sharing network, e.g., buying a (suitable) car allows one to immediately join a 'Uber'-like network of cars, or buying a mobile phone and enabling sharing of its resources via peer-to-peer connectivity enables it to become part of a distributed localised supercomputer consisting of mobile nodes.

1.2.7 Collective Computing

Collective computing proposed by Abowd in [1] integrates the cloud, the crowd and the shroud, defining a new era of 'cooperation between humans and computing that enhances both computational capabilities and the human experience.' The shroud refers to the 'layer of digital technology that connects the physical properties of people, places and things to the digital domain'. It is mentioned about collective computing that

A key feature of collective computing is a blurring of the distinction between the human and a computational element; we no longer need concern ourselves with whether an answer

---

[24]http://sloanreview.mit.edu/article/crowd-based-capitalism-empowering-entrepreneurs-in-the-sharing-economy/.

comes from a collection of computational elements or humans, or both. In fact, correct harnessing of the crowd can help machines communicate with human intelligence as efficiently as they communicate with other machines.

The notion of machine to machine connection reminds us of the device mesh and the treatment of computational elements as human or machine provides a generalised abstraction over the notion of a 'processor' in a crowd computer (which combines human and his/her mobile device(s)).

1.2.8 Swarm Dynamics

Swarm systems [2] are characterised by collective behaviours achieved in a decentralised, self-organised manner, whether artificial, as in swarm robotic systems, or natural, as in bird flocking and termite behaviours. In mobile computing, ideas from swarm systems can be explored, e.g., to coordinate crowds of devices, to steer crowds of people or for cars cooperating. We explore this idea later in the book.

1.3   Five Ideas and This Book

In this book, we will discuss five perspectives on how the above technology trends are coming together, in the following chapters.

The next chapter presents the notion of ***crowd+cloud machines***, where crowds of devices and people are pooled together to use and provide resources, often supported by cloud infrastructure.

Chapter 3 discusses cooperation in the IoT, in particular, emphasising how cooperation is facilitated and should be exploited, building on the connectivity increasingly available for things, which we call ***extreme cooperation***, giving examples from Intelligent Transport Systems involving vehicle-to-vehicle communication, crowd steering, as well as the notion of device ecologies.

Chapter 4 proposes the notion of ***scalable context-awareness***, where context-aware computing and systems can be taken further.

Chapter 5 discusses ***drone services*** in more detail, including ideas such as drones-as-a-service and how swarms of drones can help service end-users, perhaps operated by commercial providers.

Chapter 6 discusses the idea of ***social links in mobile crowds***, which discusses favour links and networks in crowds, and social networking among people and things and how that might be automated.

Chapter 7 concludes with future directions.

References

1. Gregory D. Abowd. Beyond weiser: From ubiquitous to collective computing. *Computer*, 49(1):17–23, 2016.
2. Roland Bouffanais. *Design and Control of Swarm Dynamics*. Springer, 2016.
3. Abhishek Chandra, Jon Weissman, and Benjamin Heintz. Decentralized edge clouds. *IEEE Internet Computing*, 17(5):70–73, September 2013.
4. Robin Chase. *Peers Inc: How People and Platforms Are Inventing the Collaborative Economy and Reinventing Capitalism*. PublicAffairs, 2015.
5. Cen Chen, Shih-Fen Cheng, Aldy Gunawan, and Archan Misra. Traccs: Trajectory-aware coordinated urban crowd-sourcing. In *Conference on Human Computation and Crowdsourcing (HCOMP?14)*, Pittsburgh, USA, November 2014.
6. Dingxiong Deng, Cyrus Shahabi, Ugur Demiryurek, and Linhong Zhu. Task selection in spatial crowdsourcing from worker's perspective. *Geoinformatica*, 20(3):529–568, July 2016.
7. Nathan Eagle. Txteagle: Mobile crowdsourcing. In *Proceedings of the 3rd International Conference on Internationalization, Design and Global Development: Held As Part of HCI International 2009*, IDGD '09, pages 447–456, Berlin, Heidelberg, 2009. Springer-Verlag.
8. Niroshinie Fernando, Seng W. Loke, and Wenny Rahayu. Mobile cloud computing: a Survey. *Future Gener. Comput. Syst.*, 29(1):84–106, January 2013.
9. Frank H.P. Fitzek and Marcos D. Katz. *Mobile Clouds: Exploiting Distributed Resources in Wireless, Mobile and Social Networks*. Wiley Publishing, 1st edition, 2014.
10. R.K. Ganti, Fan Ye, and Hui Lei. Mobile crowdsensing: current state and future challenges. *Communications Magazine, IEEE*, 49(11):32–39, November 2011.
11. Neil Gershenfeld. *When Things Start to Think*. Henry Holt and Co., Inc., New York, NY, USA, 1999.
12. Aakar Gupta, William Thies, Edward Cutrell, and Ravin Balakrishnan. mclerk: enabling mobile crowdsourcing in developing regions. In *Proceedings of the SIGCHI Conference on Human Factors in Computing Systems*, pages 1843–1852, Austin, Texas, USA., May 5-10 2012. ACM.
13. Jeff Howe. The rise of crowdsourcing. *Wired magazine*, 14(6):1–4, 2006.
14. Mike Kuniavsky. *Smart Things: Ubiquitous Computing User Experience Design: Ubiquitous Computing User Experience Design*. Morgan Kaufmann Publishers Inc., San Francisco, CA, USA, 2010.
15. G. Li, J. Wang, Y. Zheng, and M. J. Franklin. Crowdsourced data management: A survey. *IEEE Transactions on Knowledge and Data Engineering*, 28(9):2296–2319, Sept 2016.
16. Adam Marcus and Aditya Parameswaran. Crowdsourced data management: Industry and academic perspectives. *Found. Trends databases*, 6(1-2):1–161, December 2015.
17. Viktor Mayer-Schnberger. *Big Data: A Revolution That Will Transform How We Live, Work and Think. Viktor Mayer-Schnberger and Kenneth Cukier*. John Murray Publishers, UK, 2013.
18. Pietro Michelucci and Janis L. Dickinson. The power of crowds. *Science*, 351(6268):32–33, 2015.
19. Prayag Narula, Philipp Gutheim, David Rolnitzky, Anand Kulkarni, and Bjoern Hartmann. Mobileworks: A mobile crowdsourcing platform for workers at the bottom of the pyramid. *Human Computation*, 11:11, 2011.
20. J. Ren, Y. Zhang, K. Zhang, and X. Shen. Exploiting mobile crowdsourcing for pervasive cloud services: challenges and solutions. *Communications Magazine, IEEE*, 53(3):98–105, March 2015.
21. David Rose. *Enchanted Objects: Innovation, Design, and the Future of Technology*. Simon and Schuster, 2014.
22. M. Satyanarayanan, P. Bahl, R. Caceres, and N. Davies. The case for vm-based cloudlets in mobile computing. *IEEE Pervasive Computing*, 8(4):14–23, Oct 2009.
23. Daniel Sui, Sarah Elwood, and Michael Goodchild, editors. *Crowdsourcing Geographic Knowledge: Volunteered Geographic Information (VGI) in Theory and Practice*. Springer, 2013.

24. Arun Sundararajan. *The Sharing Economy: The End of Employment and the Rise of Crowd-Based Capitalism*. MIT Press, 2016.
25. Hien To, Cyrus Shahabi, and Leyla Kazemi.   A server-assigned spatial crowdsourcing framework. *ACM Trans. Spatial Algorithms Syst.*, 1(1):2:1–2:28, July 2015.
26. Umair ul Hassan and Edward Curry. Efficient task assignment for spatial crowdsourcing. *Expert Syst. Appl.*, 58(C):36–56, October 2016.
27. Tingxin Yan, Matt Marzilli, Ryan Holmes, Deepak Ganesan, and Mark Corner. mcrowd: A platform for mobile crowdsourcing. In *Proceedings of the 7th ACM Conference on Embedded Networked Sensor Systems*, SenSys '09, pages 347–348, New York, NY, USA, 2009. ACM.

Chapter 2
Crowd+Cloud Machines

2.1 Combining Crowd and Cloud Computing

Computing today and the future could involve tens to thousands (to millions to trillions [23]) of (mobile/stationary) nodes that can cooperate in new ways, in order to provide new capabilities and applications, from massive context-awareness to new distributed computational platforms, forming the cloud or supported by large-scale (or data-centre scale) cloud computing resources (which we call the greater Cloud).

In this chapter, we review several examples how (machine and human) resources of a (mobile) *crowd* of people with separately owned devices can be pooled together and combined with a *cloud* computing mediating platform to form a type of crowd-powered system, or what we roughly call a *crowd+cloud machine*, to emphasise this combination between the two.

In the following sections, we first review types of mobile clouds, and then consider a range of examples of crowd+cloud machines:

- crowd+cloud machines that constitute supercomputers formed out of a loosely organised crowd of mobile devices;
- crowd+cloud machines that use the crowd for decentralised spatial computations;
- crowd+cloud machines that crowdsource to search for regions of certain properties;
- crowd+cloud machines that crowdsense and recognise group activities;
- crowd+cloud machines that bridge people with disabilities and workers who can provide real-time help; and
- crowd+cloud machines for annotating the physical world.

© The Author(s) 2017   11
S.W. Loke, *Crowd-Powered Mobile Computing and Smart Things*, SpringerBriefs
in Computer Science, DOI 10.1007/978-3-319-54436-6_2

2.2   Types of Mobile Clouds

We first briefly look at a range of arrangements of computational resources or
devices, which have been termed *mobile clouds*.

- *Personal clouds*: with multiple devices on a person (e.g., health sensors, smartphone, smartwatch, smart shoes, smart jewellery and so on) and the widespread
use of short range networking technologies like Bluetooth, one could pool
together devices worn/carried by a user to form a cloud of personal resources
that mobile apps can use. For example, mobile health apps can use specialised
sensors for blood pressure, heart rate, and oxygen levels, and smart device based
physical activity tracking as well as environmental sensing (e.g., for pollutants)
and analyse the data in real-time to provide recommendations for the user, or
upload data to the greater Cloud for storage and longer term analysis.
- *Vehicular clouds*: it has been proposed that vehicle-to-vehicle (v2v) networking
or VANETS will link vehicles enabling sharing of vehicle sensor data and
management of vehicle fleets for safety, content sharing (e.g., multimedia
downloading and usage) and even virtualisation over the vehicular cloud to
do data mining [11, 18]. Particular vehicles might play the role of a leader
to coordinate the formation of a vehicular cloud with a (potentially changing)
collection of vehicles. Cars at a long term parking lot such as an airport or
a shopping centre might form a vehicular cloud but with dynamically varying
resources as cars leave or come to the car park [2]. Longer range connectivity
such as 4/5G can complement the shorter range v2v networks perhaps filling in
gaps or to access remote resources.
- *Cloudlets,*[1] *Edge Clouds, and Fog Computing*: intermediary cloud servers (or
cloudlets) between mobile users and the greater Cloud can be employed yielding
reduced latency for mobile user applications (compared to data going to and from
the remote cloud), improved security since data stays within particular geographical boundaries, and cloud outages can be masked via the edge cloudlet [32, 33].
Federation of such cloudlets can provide video analytics [34] for crowdsourced
videos (e.g., using cameras on people or cars), useful in applications such as
marketing and advertising, locating missing persons or property, and public
safety survey (e.g., to monitor infrastructure such as damaged sidewalks, icy road
surfaces and potholes)—a hierarchically structured system could pool multiple
cloudlets together.
- *Mobile crowd clouds/mobile device clouds*: such clouds are formed mainly by
pooling together devices from a nearby (say several metres or up to tens of
metres away, via Bluetooth or WiFi-Direct) crowd of people. Fitzek and Katz [9]
provided a range of ways in which a set of devices can combine resources. For
example, a set of smartphones can share loudspeakers to create 'social stereo and
3D social sound', providing amplification when, for example, all the phones play

---

[1]http://elijah.cs.cmu.edu.

the same sound files in a synchronised manner—while only a few devices are illustrated, the idea of scaling this to hundreds of devices at a rally could mean they act collectively as one powerful speaker. Or as mentioned earlier, imagine a crowd of people at a stadium watching a football match and imagine people taking pictures of the match from where they are and sharing them around— people can then have pictures and videos of the game from multiple perspectives, perhaps stitched together to form complex (almost) 3D footage of the game. Another idea is that the displays of a set of devices such as smartphones and tablets can be tiled together to form a large screen (though the inter-device boundaries cannot be removed).

* *Rethinking the Cloud for the IoT*: all the above are variations from the traditional model of cloud computing with huge building-sized array of computers and storage, towards more local pools of resources, still elastic and expandable to neighbouring clouds, and generalising on the notion of a resource that can be combined (going beyond CPU, memory and storage, to other kinds of capabilities, including sensing, connectivity, display, specialised processing functions such as video processing, analytics, positioning, media output, and so on). With the Internet of Things, we can consider multi-clouds of things, some of which are homogeneous, e.g., all the cameras in a building or along a street networked to work together, and some of which are heterogeneous, or different devices required to monitor and generate a comprehensive health report for a person. One could also think of specialised IoT clouds (1) for the elderly or the disabled (formed by the personal cloud on the individual and the local cloud in the home), (2) for streaming media such as video, (3) for agriculture, (4) for health/patient care, (5) for data mining, and (6) for a kid's room.

  Smart things might themselves also crowdsource—e.g., if a device does not understand the user, it might ask the crowd to help it understand and answer the user, and smart things might also *cloudsource*, i.e., look for cloud resources when it is unable to perform a user specified task.

In the next section, we describe a system for integrating a (dynamically varying) crowd of mobile devices for machine and human processing.

2.3 Characteristics of Crowd+Cloud Machines: The Case of Honeybee and Multi-Layered Honeybee

We consider a *crowd+cloud machine* as a 'computer' formed by a crowd of people, each with their own (network of) devices (e.g., smartphone, smartwatch, smartcar, smart-*) , inter-network-able to each other, providing human and machine computational capabilities, supported by the greater Cloud, and with varying boundaries and composition, characterised by

- *distributed ownership*: members (devices and their owners) of this crowd+cloud machine remain owned and largely administered separately by individuals, through the devices participate in a collective—the devices maintain a dual role: (1) as a participant in this crowd+cloud machine providing some of its resources in doing so, and (2) as a personal device used by its owner only;
- *heterogeneous devices*: members of the crowd-cloud machine are heterogeneous, some devices are more powerful than others and some have resources that others do not;
- *dynamic environment*: members of a crowd-cloud machine are coming and going, joining and leaving the machine, and so after a time, it could be that the set of devices that the machine consists of is an entirely or mostly different set from what it started with, even if the function the machine is performing has not changed;
- *ad hoc and opportunistic*: the crowd+cloud machine can be formed ad hoc for a specific application or opportunistically, e.g., when (the right number of) devices happen to be connect-able at the same time, the machine should then form and start working, or if a new device comes close enough with suitable resources, the machine could involve it, extending its capabilities;
- *context-aware*: the crowd-cloud machine needs to be sensitive to nearby devices and their state and to adapt accordingly;
- *localised*: closed proximity (or within the vicinity of each other) facilitates inter-action, networking possibilities, high latency connectivity, and collaborations;
- *minimal assumed knowledge of devices/resources*: for devices to work together, they should not need to know too much information about each other, i.e., the less assumed knowledge of each other, the easier the interaction mechanism;
- *can combine human and machine computation capabilities*: while the idea is to pool together compute power and memory resources for machine computations, there are tasks that could be collaboratively performed not just with machine resources but with human input, e.g., in an application to search for someone or something (say a type of plant, bird, car, or some object), humans could use their cameras to point at particular crowds or areas and to walk around, or to find objects that fit a given description, but leave it to the image processing capabilities on the phone to pick out and identify potential required targets;
- *incentive-driven*: some incentive mechanism, be it monetary payment schemes, a point system, or a favour exchange mechanism is required for cooperation to happen, and incentives are required to maintain cooperation over time;
- *energy-aware (resource-aware)*: the task collaboratively being performed by a crowd-cloud machine needs to be aware of the resources it uses, not only in adapting or for metering, but so that optimisation to improve efficiency can take place;
- *elasticity*: as mentioned, the crowd+cloud machine should be opportunistic, but if there is a need to utilise more resources, e.g., it was found that more resources are needed than anticipated, the machine should be able to connect to the greater Cloud or other neighbouring crowd-cloud machines and collaborate.

There are different degrees to which one can possess the above characteristics.

As an example with an emphasis on computational resources, we briefly describe a framework called Honeybee [7, 8, 21] that maintains the above characteristics in a limited way. Honeybee is a mobile middleware, where a group of devices with Honeybee installed can work together on a heavy computation task, broken down into a series of small tasks or jobs. Honeybee uses a distributed work stealing algorithm for automatic load balancing so that heterogenous devices can effectively work together—more powerful devices can 'steal' non-started jobs from slower devices. Honeybee also uses device and resource discovery (as built into Bluetooth and WiFi-Direct) to look for new devices that could be integrated into the collective. Devices can leave and this is detected and its jobs can then be done by other devices. It was shown that speedups of up to four can be obtained with seven devices working together on a face detection problem over a sizeable collection of photos.

Also, perhaps somewhat surprisingly, a device that delegates its work to other (especially more powerful) devices can also save energy even if it has to transmit data to other devices for processing, especially when the short range device-to-device networking technology is adequately energy-efficient. Power savings can be had not just from offloading computations—other work [5] has noted that the power required to transfer data wirelessly from a wearable device to a smartphone (to be stored) can be less than that would be needed for storing the data in flash memory on the wearable device itself.

Not only can a group of workers work for a delegator, workers themselves can delegate tasks to other workers, hence, achieving elasticity when needed; when there are resources near-by, a tree of collaborating devices can be assembled ad hoc for a task and then disassembled when the task is completed. Figure 2.1 illustrates Honeybee with the two roles that devices can play. A task is divided into a set of jobs on the delegator, and the jobs are then taken up by workers according to how fast they can work, rather than the delegator assigning tasks to workers.

In an analysis done with devices in a library at a university as noted in [21], while sitting near the centre of the building, on a regular semester day, one can see that there is a crowd of 30 to over 100 devices detected per day via Bluetooth scanning over 5–15 h per day. Hence, there is virtually a supercomputer in the library surrounding a person at any time. Alternatively, consider the idea of the phenomenon of familiar strangers [36, 37, 39], including people who transit, using public transport daily on the same routes—even though they do not know each other, there is a relatively consistent group of devices surrounding users in urban environments; a perspective on this is that there is virtually a sizeable mobile 'supercomputer' comprising tens of nodes moving together with a user (or in this way, consistently stable relative to the user).

2.4 Decentralised Spatial Computing with the Crowd

A crowd+cloud machine can be formed ad hoc to compute, in a decentralised way, results which relate to the crowd itself. Duckham [6] presents a range of algorithms for nodes in a sensor network to compute spatial properties such as boundaries

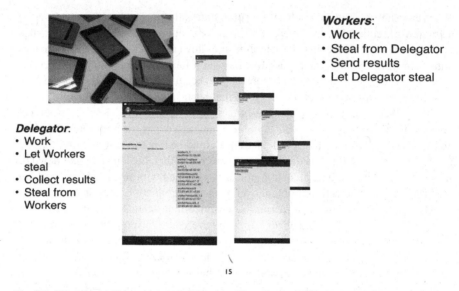

Workers:
- Work
- Steal from Delegator
- Send results
- Let Delegator steal

Delegator:
- Work
- Let Workers steal
- Collect results
- Steal from Workers

Fig. 2.1 Honeybee with workers and delegators. Note that in this implementation of Honeybee, the delegator steals from slow workers, and faster workers then steal from the delegator—we could also have faster workers steal from slower workers directly, though trading-off some delegator control

and topological relationships, as well as establishing communication networks (e.g., trees for directed diffusion style communication). Similar algorithms with feedback from the crowd can be applied to find the boundaries of a crowd of people and whether more than a certain number of people are present, say at a rally or a large concert. To take an example from Duckham,[2] the following algorithm determines if there are more than 1000 people in the crowd.

> An individual fan could start by placing a tally mark on a piece of paper. She can then pass this paper to a randomly selected neighbor, and ask them to add another tally. The paper then continues to be passed to neighbors subject to three rules:
>
> 1. please add a tally to the paper only if you have not already done so;
> 2. check the tally to see if it contains 1000 tally marks;
> 3. if it does, shout out 'Crowd!'; if not, just pass the paper to another randomly selected neighbor.
>
> Assuming the individuals in the crowd do as they are instructed, and if the crowd is large enough, sooner or later someone will shout 'Crowd!'

The above algorithm can be viewed as basically a machine that computes the number of people in the crowd formed by the crowd itself, and the result perhaps uploaded to the Cloud. One can imagine adaptations of the algorithm above to compute the number of people in the crowd which satisfy a certain criteria C, e.g., just by changing rule 1 above to:

---

[2]http://ambientspatial.net/book/?p=55.

Fig. 2.2 Illustration of a decentralised peer-to-peer algorithm to ask the crowd, at a concert say, to determine the boundary. A query is passed around to ask if the person is in the concert or not, and eventually, the replies can be aggregated to determine where the boundary is

if you satisfy criteria C, please add a tally to the paper only if you have not already done so;

Figure 2.2 illustrates another algorithm to compute the boundaries of a crowd of people actually watching a concert [25]. As a query is passed around to ask if each person is watching the concert or not, and the replies can be aggregated to determine where the boundary is, people outside the concert crowd will reply 'no' and those inside reply 'yes', eventually revealing those at the boundary. Results could be returned along the same path that the queries were sent. Perhaps such a map of the crowd can then be used by someone to find the shortest direct path out of the crowd.

Also, those within the crowd could take a picture of themselves and upload it to a cloud server so that not only is there a rough count of the number of people at the concert, but also their identification (via a picture)—while not everyone might comply, the potential of such an algorithm exists. Instead of a picture of themselves, people could take pictures or videos of the concert and upload that to the cloud.

One can also provide such an algorithm to create more sophisticated versions of the 'Mexican wave' often seen at soccer matches. The decentralised algorithm for the Mexican wave is simple[3]:

You look to your right and see the wave approaching, accompanied by a crescendo. When it hits you, you jump up and throw your hands in the air, making whatever noise you feel apposite.

---

[3]http://news.bbc.co.uk/2/hi/8742454.stm.

Someone, of course, needs to initiate the wave. And if there are several points of initiation, multiple waves can be observed. A simple algorithm is in effect commonly used to orderly leave a building for example, after a movie or an indoor event: you look to your right or left, and when your neighbour leaves, you follow the person (follow either if both leaves) till you reach the exit. Another example is a Conway's game of life style crowd behaviour formation, e.g., you are at a concert, and suppose when at least the majority of your neighbours jump, you jump for at least 1 min afterwhich you can choose to stop anytime—the overall effect, if the rule is followed, is a type of 'flock' or swarm behaviour emerging.

Interesting algorithms can be considered for crowds of people with mobile devices, e.g., leader election and computing the convex hull of a crowd of individuals, based on work in [29], to determine the boundaries of a place or the geographical range of crowd activities.

Another idea when the nodes or the people in the crowd can move or are moving is to use a decentralised algorithm via mobile-to-mobile communication to compute the best route to leave a building in case of an emergency, there are several algorithms to do this as given in [24].

2.5   Spatial Finding with the Crowd

We consider a crowd+cloud machine to compute, given bounded resources, a spatial map of a phenomenon, e.g., to compute the best possible, given the resource constraints, map of a where parking spaces are, where noisy areas are, real 3/4/5G bandwidth and coverage, as well as where the crowds in a city are currently. In [20] is a simple algorithm for crowdsourcing such urban maps while staying within a budget (since each question asked about an area could cost money, if money was paid for people to contribute information about an area). Briefly, the problem can be stated as follows.

Assume a a large area R partitioned into n regions $\{r_1, \ldots r_n\}$. The problem is to find a set $S \subseteq R$ of at least $k \leq n$ regions, each of which evaluates to true for a given predicate F representing some criteria, i.e. $F(r) = TRUE$, for each $r \in S$. We also want to solve this problem with the lowest cost (assuming we need to pay to get a question about a region answered) and in a most efficient way (the number of rounds of questions required).

For example, we want to find at least k regions with available car parking spaces, and can divide a large area into a set of regions, about which we can then ask the crowd about, but each time we ask the crowd about a region, we assume that we incur a cost. Another example is to find a not-so-crowded cafe and can issue a query to find at least k regions with a not-so-crowded cafe, answers being given by people near or within the region. A third example is to find a high bandwidth (WiFi, 4G or otherwise) region.

Figure 2.3 illustrates five regions being queried (on the left) to reveal the results (on the right). After the five queries, and suppose there is enough in the budget to

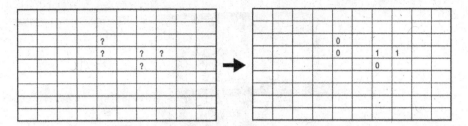

Fig. 2.3 Illustration of querying five regions to determine if the region satisfies a given property; the result is on the *right side*, where 0 means the region does not satisfy the property and 1 means it does

ask about more regions, a question is which other regions should one ask about? Note that if the phenomenon or property we are trying to find out about tends to be a clustering phenomenon so that, if one region satisfies the property, adjacent regions are likely also to satisfy the property, then a heuristic to determine which other regions to ask about would be to ask about regions which are surrounded by more 1s than 0s.

Figure 2.4 shows a coverage map and the corresponding fully disclosed idealised version. There are 7931 regions shown in the idealised version, i.e., to achieve full disclosure as in the figure, it would cost 7931 questions. This is likely too expensive and suppose there is budget for 100 questions, which of the 7931 regions would one ask about to maximise the chance of finding 1 (instead of 0) areas? For clustering phenomena such as bandwidth, a heuristic is to ask about areas adjacent to areas already found to be 1, but this could sacrifice finding new 1 clusters—hence, there is a classical exploration-exploitation trade-off. A range of heuristics can be examined for this purpose.

Other work in [38] used spatial regression techniques to learn spatial phenomena (e.g., radiation maps) represented as a continuous function from locations to values from possibly untrustworthy and noisy crowdsourced inputs, assuming the phenomenon can be modelled as such.

2.6   CAROMM and GroupSense: Crowdsensing and Crowd Activity Recognition

Crowdsensing has been explored for many years [13] and has a range of applications including crowdsourcing to detect and identify traffic regulators, traffic lights, and stop signs [14], traffic densities,[4] and in emergency evacuation scenarios [15] as well as crowd management [10] and understanding crowd behaviour in events

---

[4]See   http://www.hh.se/download/18.c3a9c5b12ba24af5f580001878/1341267487930/WWVCJai erGPart1okt10.pdf, https://www.ncbi.nlm.nih.gov/pubmed/26761013.

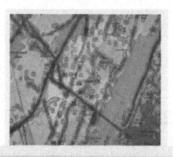

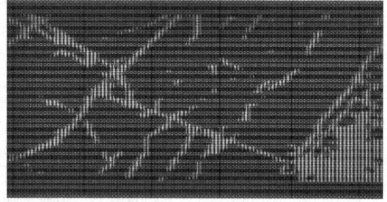

Fig. 2.4 Illustration of a 3G/4G coverage map—in the *top figure*, the *darker regions* show the coverage; the *figure below* is a discretised version where the entire area is partitioned into 103 × 77(=7931) regions—the *lighter regions* are the 1s and the *darker regions* are the 0s, corresponding to the bandwidth map above. The figure shows the fully disclosed map but the idea is that it is too expensive to ask about all 7931 regions so that some heuristic is needed for selecting the small number of regions to ask about

such as a music festival [16]. As mentioned, especially for transportation based applications, it is not only user's personal mobile devices that can be used for crowdsensing but also vehicles (e.g., [19, 40]). Such crowdsourcing can provide insightful situation awareness but issues of privacy should be taken into account [26]. A number of frameworks have emerged (e.g., [28, 30, 35]) to allow efficient and effective crowdsensing, integrating the use of crowd mobile devices as well as cloud platforms.

Not just simple sensing of the behaviour of individuals in the crowd, one can also attempt to sense and infer the activity of the crowd as a whole. There have been tremendous work in activity recognition and group activity recognition using computer vision techniques, but recent work on group activity recognition attempts to use non-image processing techniques and rely only on sensors on mobile devices and objects [3, 12, 30].

A crowd-cloud machine can be built which tracks the activity of a group or a crowd of individuals (perhaps opt-in by the individuals) and uses that to provide new mobile services to the individuals or to the group as a whole—e.g., one is providing more information beyond that of individual activity recognition, i.e., the system does not only know if a person is walking but that the person is walking together with some others (who are also walking). A tour group, a fitness group, a sports group, an expedition, or a bush-walking group can have their activities tracked via a cloud assisted platform, using mobile sensing readily available from the individuals' mobile devices, and reasoning with and aggregating the mobile sensor data from the group via cloud analytics.

2.7 Crowd+Cloud Machines to Assist People with Disabilities

While many of the above applications use mobile computational resources and sensing, there is opportunity for mobile crowdsourcing to involve users, not only on the worker or service side as in the spatial crowdsourcing mentioned earlier but also on the client side for whom the workers are doing their tasks. An example of this is the system called VizWiz[5] for blind users to crowdsource and receive quick answers to questions about their surroundings, and others reviewed in [4]. Chorus:View [17] is a system that assists users with workers getting into a continuous conversation with the user about a video stream from the user's mobile device. Exploring crowd help to aid the visually impaired navigate along paths was also explored in [22], where a video stream from a client's mobile device is sent to a cloud platform and then forwarded to workers to be viewed so that they can then advise the user on his/her navigation. Apart from human workers, image processing algorithms can also be used to aid visually impaired clients.

A crowd-cloud machine can be built to help the disabled navigate places but issues of responsibility and risks remain open as to the extent and quality of help that can be given by remote workers.

2.8 Physical Annotation Systems

Physical annotation systems are systems (typically involving a mobile app for users as in [1]) for users to leave notes (which can be videos, photos, or audio, and not just text) or information that can be associated with particular objects or places, similar to how one might mark up a piece of text and link it with notes. The idea is that a layer (or multiple layers) of annotations can be added to the physical world, and stored in the cloud. The association can be done by associating an RFID tag of an object or place with the information in a database, but other forms of object or place identification can be used such as GPS coordinates and even just pictures of

---

[5]http://vizwiz.org.

objects. This enables not only other users to retrieve and read information (e.g., via an augmented reality style app) about a particular place or object that other users left for that place or object, but also allows robots and drones to obtain additional information about objects and places in the real world.

Humans (via crowdsourcing) and machines with aggregation and filtering algorithms might be needed to moderate the annotations. The system of mobile apps (for users to write annotations and link them to objects and places or to read annotations) and client software (e.g., on robots and drones to manipulate annotations) together with the cloud server (for storage and management of such information) constitutes the crowd+cloud machine for physical annotation.

2.9   Summary

We have reviewed a range of crowd+cloud machines, which are essentially distributed systems formed by mobile resources (often with human involvement) and cloud services. The different crowd+cloud machines can be combined. For example, the physical annotation systems can be combined with a machine for helping the disabled, so that the annotations can be read to clients who are visually impaired to provide more information about a place or to help people with memory loss [27]. When processing images or videos, resources from a Honeybee like pool of devices might be used, e.g., to identify faces in video streams used in a crowd-cloud machine to help the disabled when workers are not available or reliable. In the absence of human workers, crowd activity recognition can be used to automatically annotate what a group of people in front of a blind person is doing, and then this is read to the person. Crowdsensing can be used to obtain additional contextual information about the surroundings to predict available resources for upcoming jobs to be done via Honeybee-like computations. There are many interesting systems that could be considered crowd+cloud machines which we did not review here—an interesting example is *crowd physics* [31] which considers using people to help deliver packages. Uber delivery[6] employs people to deliver food and goods for relatively small payments. Such a notion, if viewed as a system, albeit an open one, can be viewed as a crowd+cloud machine for deliveries.

References

1. Ahmad A. Alzahrani, Seng W. Loke, and Hongen Lu. An advanced location-aware physical annotation system: From models to implementation. *J. Ambient Intell. Smart Environ.*, 6(1): 71–91, January 2014.

---

[6]https://www.uber.com/deliver

2. S. Arif, S. Olariu, J. Wang, G. Yan, W. Yang, and I. Khalil. Datacenter at the airport: Reasoning about time-dependent parking lot occupancy. *IEEE Transactions on Parallel and Distributed Systems*, 23(11):2067–2080, Nov 2012.

3. Amin Bakhshandehabkenar, Seng W. Loke, and J. Wenny Rahayu. A framework for continuous group activity recognition using mobile devices: Concept and experimentation. In *IEEE 15th International Conference on Mobile Data Management, MDM 2014, Brisbane, Australia, July 14-18, 2014 - Volume 2*, pages 23–26, 2014.

4. Erin Brady and Jeffrey P. Bigham. Crowdsourcing accessibility: Human-powered access technologies. *Found. Trends Hum.-Comput. Interact.*, 8(4):273–372, November 2015.

5. R. Chandra, S. Hodges, A. Badam, and J. Huang. Offloading to improve the battery life of mobile devices. *IEEE Pervasive Computing*, 15(4):5–9, Oct 2016.

6. Matt Duckham. *Decentralized Spatial Computing: Foundations of Geosensor Networks*. Springer Publishing Company, Incorporated, 2012.

7. N. Fernando, Seng W. Loke, and W. Rahayu. Computing with nearby mobile devices: a work sharing algorithm for mobile edge-clouds. *IEEE Transactions on Cloud Computing*, PP(99): 1–1, 2016.

8. Niroshinie Fernando, Seng W. Loke, and Wenny Rahayu. Honeybee: A Programming Framework for Mobile Crowd Computing. In *Proc. of the 9th Int. Conf. on Mobile and Ubiquitous Systems: Comp., Netw. and Serv. (MobiQuitous)*, pages 224–236, 2012.

9. Frank H.P. Fitzek and Marcos D. Katz. *Mobile Clouds: Exploiting Distributed Resources in Wireless, Mobile and Social Networks*. Wiley Publishing, 1st edition, 2014.

10. Tobias Franke, Paul Lukowicz, Martin Wirz, and Eve Mitleton-Kelly. Participatory sensing and crowd management in public spaces. In *Proceeding of the 11th Annual International Conference on Mobile Systems, Applications, and Services*, MobiSys '13, pages 485–486, New York, NY, USA, 2013. ACM.

11. M. Gerla, E. K. Lee, G. Pau, and U. Lee. Internet of vehicles: From intelligent grid to autonomous cars and vehicular clouds. In *Internet of Things (WF-IoT), 2014 IEEE World Forum on*, pages 241–246, March 2014.

12. Dawud Gordon. *Group Activity Recognition Using Wearable Sensing Devices*. PhD thesis, Karlsruhe Institute of Technology, 2014.

13. Bin Guo, Zhu Wang, Zhiwen Yu, Yu Wang, Neil Y. Yen, Runhe Huang, and Xingshe Zhou. Mobile crowd sensing and computing: The review of an emerging human-powered sensing paradigm. *ACM Comput. Surv.*, 48(1):7:1–7:31, August 2015.

14. Shaohan Hu, Lu Su, Hengchang Liu, Hongyan Wang, and Tarek F. Abdelzaher. Smartroad: Smartphone-based crowd sensing for traffic regulator detection and identification. *ACM Trans. Sen. Netw.*, 11(4):55:1–55:27, July 2015.

15. Azhar Mohd Ibrahim, Ibrahim Venkat, K. G. Subramanian, Ahamad Tajudin Khader, and Philippe De Wilde. Intelligent evacuation management systems: A review. *ACM Trans. Intell. Syst. Technol.*, 7(3):36:1–36:27, February 2016.

16. Jakob Eg Larsen, Piotr Sapiezynski, Arkadiusz Stopczynski, Morten Mørup, and Rasmus Theodorsen. Crowds, bluetooth, and rock'n'roll: Understanding music festival participant behavior. In *Proceedings of the 1st ACM International Workshop on Personal Data Meets Distributed Multimedia*, PDM '13, pages 11–18, New York, NY, USA, 2013. ACM.

17. Walter S. Lasecki, Phyo Thiha, Yu Zhong, Erin Brady, and Jeffrey P. Bigham. Answering visual questions with conversational crowd assistants. In *Proceedings of the 15th International ACM SIGACCESS Conference on Computers and Accessibility*, ASSETS '13, pages 18:1–18:8, New York, NY, USA, 2013. ACM.

18. E. Lee, E. K. Lee, M. Gerla, and S. Y. Oh. Vehicular cloud networking: architecture and design principles. *IEEE Communications Magazine*, 52(2):148–155, February 2014.

19. Yazhi Liu, Jianwei Niu, and Xiting Liu. Comprehensive tempo-spatial data collection in crowd sensing using a heterogeneous sensing vehicle selection method. *Personal Ubiquitous Comput.*, 20(3):397–411, June 2016.

20. Seng W. Loke. Heuristics for spatial finding using iterative mobile crowdsourcing. *Hum.-centric Comput. Inf. Sci.*, 6(1):61:1–61:31, December 2016.

21. Seng W. Loke, Keegan Napier, Abdulaziz Alali, Niroshinie Fernando, and Wenny Rahayu. Mobile computations with surrounding devices: Proximity sensing and multilayered work stealing. *ACM Trans. Embed. Comput. Syst.*, 14(2):22:1–22:25, February 2015.
22. Seng W. Loke and Batni Prabhanjan. Guidemate: a crowd-powered system to assist the disabled. In *Proceedings of the 2015 ACM International Joint Conference on Pervasive and Ubiquitous Computing and Proceedings of the 2015 ACM International Symposium on Wearable Computers, UbiComp/ISWC Adjunct 2015, Osaka, Japan, September 7-11, 2015*, pages 1379–1384, 2015.
23. Peter Lucas, Joe Ballay, and Mickey McManus. *Trillions: Thriving in the Emerging Information Ecology*. Wiley Publishing, 1st edition, 2012.
24. Sabrina Merkel. *Building Evacuation with Mobile Devices*. KIT Scientific Publishing, 2014. Available at http://www.ksp.kit.edu/9783731502074.
25. J. Phuttharak and S. W. Loke. Towards declarative programming for mobile crowdsourcing: P2p aspects. In *2014 IEEE 15th International Conference on Mobile Data Management*, volume 2, pages 61–66, July 2014.
26. Layla Pournajaf, Daniel A. Garcia-Ulloa, Li Xiong, and Vaidy Sunderam. Participant privacy in mobile crowd sensing task management: A survey of methods and challenges. *SIGMOD Rec.*, 44(4):23–34, May 2016.
27. Eduardo Quintana and Jesus Favela. Augmented reality annotations to assist persons with alzheimers and their caregivers. *Personal and Ubiquitous Computing*, 17(6):1105–1116, 2013.
28. Moo-Ryong Ra, Bin Liu, Tom F. La Porta, and Ramesh Govindan. Medusa: A programming framework for crowd-sensing applications. In *Proceedings of the 10th International Conference on Mobile Systems, Applications, and Services*, MobiSys '12, pages 337–350, New York, NY, USA, 2012. ACM.
29. Sergio Rajsbaum and Jorge Urrutia. Some problems in distributed computational geometry. *Theoretical Computer Science*, 412(41):5760–5770, 2011.
30. Haggai Roitman, Jonathan Mamou, Sameep Mehta, Aharon Satt, and L.V. Subramaniam. Harnessing the crowds for smart city sensing. In *Proceedings of the 1st International Workshop on Multimodal Crowd Sensing*, CrowdSens '12, pages 17–18, New York, NY, USA, 2012. ACM.
31. Adam Sadilek, John Krumm, and Eric Horvitz. Crowdphysics: Planned and opportunistic crowdsourcing for physical tasks. In *Proceedings of the Seventh International Conference on Weblogs and Social Media, ICWSM 2013, Cambridge, Massachusetts, USA, July 8-11, 2013.*, 2013.
32. M. Satyanarayanan. The emergence of edge computing. *Computer*, 50(1):30–39, Jan 2017.
33. M. Satyanarayanan, P. Bahl, R. Caceres, and N. Davies. The case for vm-based cloudlets in mobile computing. *IEEE Pervasive Computing*, 8(4):14–23, Oct 2009.
34. M. Satyanarayanan, P. Simoens, Y. Xiao, P. Pillai, Z. Chen, K. Ha, W. Hu, and B. Amos. Edge analytics in the internet of things. *IEEE Pervasive Computing*, 14(2):24–31, Apr 2015.
35. W. Sherchan, P. P. Jayaraman, S. Krishnaswamy, A. Zaslavsky, S. Loke, and A. Sinha. Using on-the-move mining for mobile crowdsensing. In *2012 IEEE 13th International Conference on Mobile Data Management*, pages 115–124, July 2012.
36. Lijun Sun, Kay W. Axhausen, Der-Horng Lee, and Xianfeng Huang. Understanding metropolitan patterns of daily encounters. *Proceedings of the National Academy of Sciences*, 110(34):13774–13779, 2013.
37. Jameson L. Toole, Yves-Alexandre de Montjoye, Marta C. González, and Alex (Sandy) Pentland. *Modeling and Understanding Intrinsic Characteristics of Human Mobility*, pages 15–35. Springer International Publishing, Cham, 2015.
38. Matteo Venanzi, Alex Rogers, and Nicholas R. Jennings. Crowdsourcing spatial phenomena using trust-based heteroskedastic gaussian processes. In *Proceedings of the First AAAI Conference on Human Computation and Crowdsourcing, HCOMP 2013, November 7-9, 2013, Palm Springs, CA, USA*, 2013.

39. Fusang Zhang, Beihong Jin, Tingjian Ge, Qiang Ji, and Yanling Cui. Who are my familiar strangers? revealing hidden friend relations and common interests from smart card data. In *Proceedings of the 25th ACM International on Conference on Information and Knowledge Management*, CIKM '16, pages 619–628, New York, NY, USA, 2016. ACM.
40. Wangsheng Zhang, Guande Qi, Gang Pan, Hua Lu, Shijian Li, and Zhaohui Wu. City-scale social event detection and evaluation with taxi traces. *ACM Trans. Intell. Syst. Technol.*, 6(3):40:1–40:20, May 2015.

Chapter 3
Extreme Cooperation with Smart Things

3.1 Things Cooperating More Than Ever

The development of energy efficient long and short range networking technologies among mobile devices is enabling the device mesh mentioned in Chap. 1, between all types of mobile devices, including smart vehicles and smart drones, e.g., vehicle-to-vehicle, vehicle-to-pedestrian, pedestrian-to-pedestrian, vehicle-to-bicycle, bicycle-to-bicycle, drone-to-vehicle, drone-to-drone, drone-to-pedestrian, and so on. Over such a networked mesh of devices can be a range of different cooperation protocols, specific to particular applications, from vehicles talking to each other to improve safety and situation-awareness, to vehicles talking about the route to take in order to avoid congestion.

Cooperation among smart things in the home can be useful too—e.g., the entertainment devices from the television to the sound system as well as lighting (and any intrusive devices disabled) cooperate to give the user the best home cinema experience. For IoT, meaningful links can be formed among smart things to clarify how they can work together but also how they should work together. For security reasons, one could monitor and enforce only certain kinds of interaction among devices, ruling out other spurious or unauthorised interactions.

Human cooperation can also be facilitated via direct interactions among smart things and human cooperation is involved for sharing things.

This chapter explores examples of cooperation among smart things, and how that could relate to human cooperation, the first relating to Intelligent Transport Systems and smart vehicles, namely, cooperation among vehicles to avoid congestion and cooperation among vehicles to resolve contention for parking spaces, and so on, the second concerns cooperation among smart things according to meaningful relationships and workflows, the third involves large crowd cooperation in large events or emergency situations (which relates to the crowd-cloud machine for decentralised spatial computations we saw earlier), and the fourth concerns cooperation for sharing things.

© The Author(s) 2017
S.W. Loke, *Crowd-Powered Mobile Computing and Smart Things*, SpringerBriefs in Computer Science, DOI 10.1007/978-3-319-54436-6_3

3.2   Vehicle-to-Vehicle Cooperation

This section explores a range of applications for vehicle-to-vehicle cooperation.

3.2.1   Benefits of Cooperation to Reduce Traffic Congestion

In general, when vehicles know which routes are congested and which are not, they could try to avoid such congestion and take less congested routes thereby saving time. But if all vehicles were to do so, say most or all shift to the less congested route, even if it seems beneficial to each car individually, the less congested route could now be just as congested, thereby reducing savings. Cooperation among cars could alleviate the problem by allowing cars to 'agree' to use separate routes and so distribute themselves out along different routes.

Another way to see how cooperation could benefit vehicles is to revisit Braess's Paradox (this version from [10]). Consider a road network shown in Fig. 3.1a, where there are cars wanting to go from *start* to *end*, and there are two ways to get from *start* to *end*, via A or via B. The cost (say in time $t(x)$) to go through a link varies with x, where x is the fraction of cars going through that link. The constant y is a fixed travel time from A to *end* and from *start* to B, regardless of how many cars go through the *start* $\rightarrow B$ and $A \rightarrow end$ links, the cost is y. Now both routes *start* $\rightarrow A \rightarrow end$ and *start* $\rightarrow B \rightarrow end$ cost the same, namely, $x + y$. And so, when $x = 0.5$, to each car, the total cost of either route is $0.5 + y$, i.e. cars can just choose randomly, and so, on average, they would somehow likely by chance to split themselves up equally between the routes, then all cars will take time $0.5 + y$.

Now, suppose a new link is added as in Fig. 3.1b with teleportation so that its cost is 0, then there is a split at A. Suppose $x \leq 1 < y < 1.5$, then the route via the new link could take $(x + 0 + x) \leq (x + y)$, i.e., the new route *start* $\rightarrow A \rightarrow B \rightarrow end$ is potentially the fastest. Hence, each car might individually decide to take that route, and note that if they do so, each would take time $(1 + 0 + 1)$, which is $2 > (0.5 + y)$,

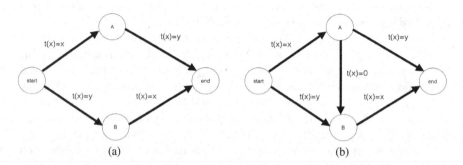

Fig. 3.1   Braess's Paradox. (a) Original network. (b) Network with new link

more than the time they would have taken if they all just ignored the new link and split themselves evenly over $start \rightarrow A \rightarrow end$ and $start \rightarrow B \rightarrow end$; but to avoid this situation, cars can cooperate so that they all agree to ignore the new link. Note that if $y \geq 1.5$, then potentially, the new route would be the best choice for all the cars.

CARAVAN [5] is a multiagent algorithm developed to exploit cooperation among vehicles but in a decentralised manner. Assuming Dedicated Short Range Communication (DSRC) type vehicle-to-vehicle networking is available, the idea is that vehicles near each other at junctions can cooperate—negotiate about the routes to take for their destinations, and decide to effectively distribute themselves out on separate routes so as not to congest on any one route. Simulations have shown that travel time savings as much as 25–35% can be obtained when cars cooperate this way. The idea is that disparate negotiations at different junctions in the area can lead to global alleviation of congestion throughout the area. Figure 3.2 shows the road network for an artificial scenario, and a Melbourne CBD scenario where cars at junctions cooperate as they go along a route from A to B (dotted circles indicate some junctions where negotiation takes place).

An MIT study using data for cities such as San Francisco and Boston revealed that cooperative route planning among vehicles (even if not necessarily using a vehicle-to-vehicle decentralised approach) can reduce congestion by as much as 30%, where congestion is measured based on the extra travel time due to traffic compared with free flow traffic.[1] Sometimes, it might not be all gain; for example, in some situations, it might be that some drivers might spend an additional few minutes to take an optional offshoot and cars might need to travel further, but as a result, overall, most drivers could gain time, some as much as 10 min.

3.2.2   Cooperating over Time

Another consideration is when, for the network in Fig. 3.1a above, let $y = 1$ for simplicity, suppose there are ten cars, and on certain days (say even days), a group of three cars are indifferent as to how long they take to get to the end than on other days, i.e., for day d, we have, for $i \in \{1, 2, 3\}$, let the benefit of getting to the end in time $(x + 1)$, from the car's perspective, be given as:

$$b_i(d) = -(d \bmod 2) * c_i(d)$$

that is, the benefit is 0 on even days regardless of the time cost; on odd days, the benefit is the negative of the total cost $c_i(d)$ of getting from start to end.

---

[1]http://spectrum.ieee.org/cars-that-think/transportation/efficiency/cooperative-route-planning-could-make-driving-slightly-less-terrible-for-everyone.

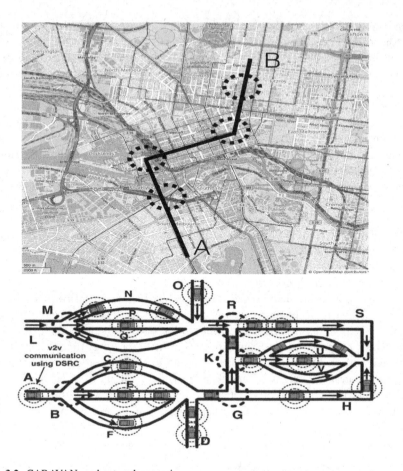

Fig. 3.2 CARAVAN road network scenarios

But it is the inverse for the rest in the other group, i.e. $j \in \{4, \ldots, 10\}$, don't care how long it takes on odd days but on even days, the benefit is the negative of the total cost, i.e. we have:

$$b_j(d) = -(1 - d \bmod 2) * c_i(d)$$

Now consider a period of 6 days (d=1, 2, 3, 4, 5, and 6). If all the cars simply decided each day to split themselves up evenly, i.e. $x = 0.5$, each day, then the total benefit of each car 1, 2, and 3 over the 6 days as

$$-(0.5 + 1) + 0 + -(0.5 + 1) + 0 + -(0.5 + 1) + 0 = -4.5$$

and for each car 4–10 over the 6 days, the benefit would be

$$0 + -(0.5 + 1) + 0 + -(0.5 + 1) + 0 + -(0.5 + 1) = -4.5$$

However, if the cars cooperated so that on odd days, cars 1, 2 and 3 all travel on one route ($x = 0.3$), and the rest on the other route, but on even days, cars split themselves up evenly as before, then we have the benefit of each car, for cars 1, 2, and 3, over the 6 days is

$$-(0.3 + 1) + 0 + -(0.3 + 1) + 0 + -(0.3 + 1) + 0 = -3.9$$

and for each car 4–10 over the 6 days the benefit would still be -4.5, so that cars 1, 2 and 3 are better off at no reduction of benefit to the rest.

Over longer periods, the cars can even better improve on their situation by cooperatively taking turns travelling alone on a route, provided there is extra benefit in getting from start to end very quickly; for example, suppose for all cars the benefit is as follows for any day, for $i \in \{1, \ldots, 10\}$:

$$b_i = \begin{cases} 10, & \text{if } c_i \leq 1.1 \\ -c_i, & \text{otherwise} \end{cases} \tag{3.1}$$

that is, whenever $c_i \leq 1.1$, there is a reward of 10. Over 10 days, distributing themselves over the two routes evenly each day, that is, each car incurring a cost of $c_i = x + y = 0.5 + 1$ each day means we have a total cost of $-1.5 * 10 = -15$ over the 10 days. But if the cars cooperated so that one car travels alone on a day and the cars take turns doing so. For example, in day 1, car 1 travels on a route alone (with cost $c_1 = 0.1 + 1 = 1.1$) and the rest travels on the other route (with cost $0.9 + 1 = 1.9$), the benefits are as follows:

$$day\ 1, car\ 1 : 10$$

$$day\ 1, car\ 2\ to\ 10 : -1.9$$

and on day 2, we have:

$$day\ 2, car\ 2 : 10$$

$$day\ 2, car\ 1,\ and\ cars\ 3\ to\ 10 : -1.9$$

and so on. Then, over 10 days, the total benefit of each is $9 * (-1.9) + 10 = -7.1$. Each car managed to reap the benefit of getting there earlier for 1 day, but need to cooperate so that the benefit is fairly experienced by all.

The idea is that with long term cooperation, there is greater benefit that can be had and fairness over the long term achieved, even if it seems unfair in each current time period, than if each decision was only to be decided now. In practice is already the idea of staggering work times, so that cars utilise the roads at different times, to distribute the traffic over time.

The notion of socially networked cars [9], where cars form a social network separate from the social network of people has been proposed, as a means of vehicles 'remembering' each other and perhaps facilitating a favour exchange or *quid pro quo* scheme. Decentralised blockchain style recording of transactions or favours made might be employed so that such favour exchange or 'brownie' points scheme might be recorded as explored for ride-sharing.[2] The work in [7] aims to provide social networking among people in nearby cars, over vehicle-to-vehicle communications, based on provided user profiles and estimated connection times for vehicle-to-vehicle links, but our idea here is social networking among the cars themselves.

3.2.3   Cooperation to Resolve Contention for Car Park Spaces

Apart from vehicle-to-vehicle cooperation for car parking and safety,[3] one could use such cooperation to resolve conflicts when contending for car park spaces, or at least to reduce congestion when cars compete for car park spaces.

Figure 3.3 shows a parking area with cars equipped with agents which can (1) help each other find car park by exchanging information each obtained, or (2) the agents can automatically negotiate with each other in case multiple cars found out that they are aiming for the same parking space—a conflict can be resolved when a car gets to know (e.g., via DSRC messages) that another car nearer to it intends to park at the space it is targeting [1].

For (1), there are issues of contention where cars might not want to share information to other cars in case this increases competition—so a car might only share information about car park spaces that it is not interested in, or after it has parked.

For (2), however, various contention resolution mechanisms can be used (e.g., by just giving in to the car nearer to the car park space in case two vehicles found out they are contending for the same space)—while this has been shown in [1] to reduce the searching time for spaces (i.e., the time-to-park), there are certainly issues of fairness to consider here, as well as issues of whether such cooperation is more or less fair compared to centralised parking allocation mechanisms.

Centralised car park space allocation can help reduce contention but at the cost of infrastructure to provide efficient, accurate and timely information about car park spaces, as well as the issue of whether cars will be happy with the recommended or assigned car park space.

---

[2]http://lazooz.org.

[3]https://www.technologyreview.com/s/534981/car-to-car-communication/.

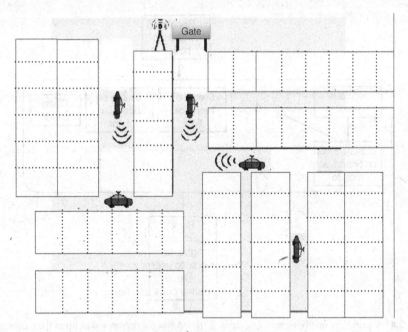

Fig. 3.3 Cooperative Parking—each vehicle is illustrated with its residing agent and the agents cooperate with other agents in resolving contention for car park spaces

3.3 Interactions and Relationships in Cooperative Living Room IoT: Device Ecologies

Devices in a living room can be coordinated to serve the human inhabitants. A simple workflow to prepare the home environment when waking up very early in the morning (while still dark) in a smart home example, based on the one in [6], is illustrated in Fig. 3.4. The idea is that a set of devices can have (perhaps formally specified) semantically meaningful relationships with one another [11], and so, not only are their interactions regulated, they are also well-defined, e.g., a device can substitute another, complement another, mediate between two devices, and enable, disable, enhance other devices, or mutually benefit each other (i.e., a symbiosis). With such relationships, such devices within a locale form a *device ecology*.

Such workflows can scale up to devices beyond the living room (or a single home) to neighbourhoods and even streets and cities. Smart things themselves might automate such cooperation. For example, the set of appliances in an entire neighbourhood cooperate to use electricity at different times to lower peak demand requirements. Peak demand requirements might mean that a power company needs to rely more on fossil fuel generation or to source electricity elsewhere at higher costs. While there are freezers that operate only when electricity is cheap provided

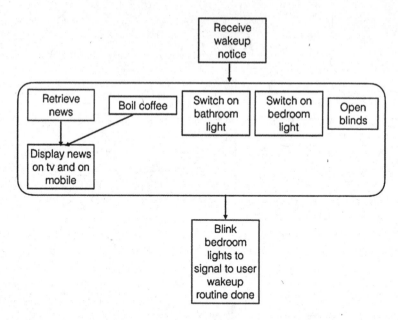

Fig. 3.4 A workflow involving multiple devices started when a person wakes up in the morning

their temperature allows this,[4] perhaps a far-fetched example is to consider all the freezers and heaters in a city cooperating to manage overall electricity demand to avoid high peak loads.

The notion of the Social Internet of Things[5] involving 'social objects' was proposed in [2] that are 'able to discover new services, start new acquaintances, exchange information, connect to external services, exploit other objects' capabilities, and collaborate toward a common goal', have 'friends' and can form their own social network. Such relationships among social objects can lead to the formation of long term relationships among such objects. Cars can form such relationships among themselves, separate from the relationships among people (or their owners) as suggested in [9]. We discuss this further in Chap. 6.

---

[4]http://www.goodnewsfinland.com/finnish-innovation-enables-home-appliances-to-avoid-electricity-demand-peaks/.

[5]http://www.social-iot.org.

3.4 Cooperation Within Large Crowds: The Case of Crowd Steering

From vehicular cooperation, we consider crowds with mobile devices cooperating to steer themselves out of an area. We consider the idea of *crowd steering*, and also the notion of mobile-based *swarm evacuation*.

In crowd steering [3], collective movements of people are tracked and movements encouraged via mobile app advice or other mechanisms such as the use of public displays, and the 'nudging' can be done via vibration, touch and sounds, and not just displayed messages. The reasons for steering the crowd might be for emergency evacuation, guided tours, safe movement of people during large rallies and concerts, regulating the use of spaces or nudging movements for commercial purposes (e.g., to encourage crowds to move through certain areas to help businesses in the areas). There is certainly a balance of control required where the crowd needs to cooperate with one another and go (or not) with the nudges they receive. In the decentralised system for swarm-based smart evacuation from a building described in [8], a range of techniques have been examined, using peer-to-peer cooperation and enabling position mapping of the people (e.g., where the crowds are) using swarm-based communication.

While such swarm evacuation can be used for people, a similar technique can be explored for vehicles needing to evacuate a place (e.g., earthquake or bush fire affected areas) in a more efficient way.

3.5 Cooperation for Sharing Things: Decentralised?

Cooperation can be mediated by a centralised or decentralised platform, though a decentralised platform removes the need for central management and control. People can be cooperating towards a 'decentralised' version of Uber-like ride-sharing where there is no one company or platform mediating ride providers and ride consumers.[6]

A platform for mediating consumers and providers has been effective for the sharing economy, from restaurant reservations such as OpenTable,[7] to people who want to share car park spaces[8] and their umbrella as mentioned earlier. As noted in [4], people with resources to rent out or share need a platform to find people who need those resources. A central platform for doing so is convenient and provides a

---

[6]See http://www.shareable.net/blog/cabby-owned-taxi-cooperatives-on-the-rise, http://www.carfreecorvallis.com/2011/12/bring-decentralized-car-sharing-to-corvallis/, https://www.inverse.com/article/13500-arcade-city-is-a-blockchain-based-ride-sharing-uber-killer, http://www.shareable.net/blog/lazooz-the-decentralized-crypto-alternative-to-uber, http://lazooz.org.

[7]http://www.opentable.com/.

[8]https://www.justpark.com.

focal point for both providers and consumers of resources and to coordinate their interactions, perhaps even providing producer and consumer vetting, safety and payment features.

However, the notion of a decentralised platform for aiding such sharing of resources is only beginning to be explored. It is not only a decentralised use of shared resources, but a decentralised coordination of how the resources are shared, and in the case of payment mechanisms, perhaps even decentralised blockchain-style recording of 'payments'. Such decentralised mechanisms might not work in all instances. For example, consider a decentralised version of OpenTable where only short range networking is allowed—e.g., suppose there is a mobile app that users install where each user's mobile device keeps track of restaurants it knows about which currently has vacancies and receives updates from the restaurants about vacancies when it comes into close proximity with them, and the app proactively shares such vacancy information with other interested mobile devices. There would be limitations in terms of the reach and timeliness of such an application—using centralised coordination with wide area networking will certainly help here. It remains to be seen if totally decentralised bitcoin-style recording of favours made can scale and fuel the sharing economy when it comes to ride-sharing or other applications.

Cooperation can still be used to resolve contention for popular resources, e.g., when several parties want to eat at a restaurant that is fully booked, can they cooperate so that some trade-off eating at the restaurant 1 day for favours to eat at another restaurant on another day? Or cross-domain trades might be carried out, e.g., one party passing on its restaurant reservation to another party in exchange for the use of a car for a certain time. While money basically efficiently mediates across such trades, can a decentralised bitcoin-style mechanism be employed to record and manage such favour exchange schemes? Such cooperation requires technology to make it efficient and easy to do so.

The idea is that if cooperation can be made easier and facilitated via some platform, there could be a greater use of cooperative mechanisms, e.g., a set of Uber-like cars can cooperate with one another to transfer passengers as efficiently as possible and with reduced user wait and travel times, all without central coordination. A set of vehicles can cooperate to provide transport for the entire journey of a crowd of people, involving efficient shared use of vehicles where possible, and all coordinated in a decentralised opportunistic manner. Smart things themselves may learn to assess their own utilisation and attempt to share themselves with others via their networks.

3.6 Correlated Equilibrium

A very simple illustration of the benefits of cooperation, inspired by game theory, can be seen in Table 3.1, where if both cars at a junction choose the same route, both are slower than if they chose different routes, and one way to guarantee different

Table 3.1 Table showing the disadvantages of both choosing the same route, compared to choosing different routes

	Car B chooses route 1	Car B chooses route 2
Car A chooses route 1	Travel times for A:30 min,B:30 min	Travel times for A:15 min,B:20 min
Car A chooses route 2	Travel times for A:20 min,B:15 min	Travel times for A:40 min,B:40 min

routes are chosen is by both cars cooperating and agreeing to go on different routes. If both chose the shorter route (i.e., route 1) individually, then both are slower than if they used different routes.

Another possibility is that some central authority at the junction tells each car to go on different routes (assigning different routes to each car) thereby reducing the travel times of both.

In the case where each car is advised of a recommended route (arrived at either via a central agency or a result from a run of a decentralised cooperative algorithm). Suppose the recommendation is either (A-1,B-2) or (A-2,B-1) with equal probability. Then, neither car would want to (or should) deviate from its given recommendation, given that it knows the other car will accept its recommendation, and we have a correlated equilibrium, with the highest advantage to both. But, as noted earlier, cooperating over time might be needed to ensure fairness, since if one of the cars is 'assigned' the shorter route all the time, it might not be happy (though if both the recommendations are given with equal probability, it should be fair in the long run).

Note, however, that some prior knowledge of traffic conditions would be required in practice for the right recommendations to be given.

3.7 Summary

The focus of this chapter has been on cooperation with smart things where we considered examples from vehicle-to-vehicle cooperation to cooperation among living room appliances. Cooperation can be a means to realise a Pareto optimal solution when without cooperation, an inferior (even if a Nash equilibrium) solution is the best that can be achieved. We also considered the idea of longer term cooperation among devices which is only beginning to be explored, and also considered cooperation involved when things (including smart cars) are to be shared, from both centralised and decentralised perspectives. The point of this chapter is that technology is beginning to enable greater possibilities for cooperation involving smart things, as well as facilitating sharing of smart things and smart things sharing resources.

References

1. A. Aliedani, S. W. Loke, A. Desai, and P. Desai. Investigating vehicle-to-vehicle communication for cooperative car parking: The copark approach. In *2016 IEEE International Smart Cities Conference (ISC2)*, pages 1–8, Sept 2016.
2. L. Atzori, A. Iera, and G. Morabito. From "smart objects" to "social objects": The next evolutionary step of the internet of things. *IEEE Communications Magazine*, 52(1):97–105, January 2014.
3. Claudio Borean, Roberta Giannantonio, Marco Mamei, Dario Mana, Andrea Sassi, and Franco Zambonelli. Urban crowd steering: An overview. In *Proceedings of the 8th International Conference on Internet and Distributed Computing Systems - Volume 9258*, IDCS 2015, pages 143–154, New York, NY, USA, 2015. Springer-Verlag New York, Inc.
4. Robin Chase. *Peers Inc: How People and Platforms Are Inventing the Collaborative Economy and Reinventing Capitalism*. PublicAffairs, 2015.
5. Prajakta Desai, Seng Wai Loke, Aniruddha Desai, and Jugdutt Singh. CARAVAN: congestion avoidance and route allocation using virtual agent negotiation. *IEEE Trans. Intelligent Transportation Systems*, 14(3):1197–1207, 2013.
6. S. W. Loke, M. E. Orlowska, S. Weerawarana, M. P. Papazoglou, J. Yang, Service-oriented device ecology workflows. In *Proceedings of the Service-Oriented Computing - ICSOC 2003: First International Conference, Trento, Italy, December 15–18*, 2003, pages 559–574, Springer, Berlin/Heidelberg. ISBN:978-3-540-24593-3, doi:10.1007/978-3-540-24593-3_38, http://dx.doi.org/10.1007/978-3-540-24593-3_38.
7. T. H. Luan, X. Shen, F. Bai, and L. Sun. Feel bored? join verse! engineering vehicular proximity social networks. *IEEE Transactions on Vehicular Technology*, 64(3):1120–1131, March 2015.
8. Sabrina Merkel. *Building Evacuation with Mobile Devices*. KIT Scientific Publishing, 2014.
9. A. Rakotonirainy and S.W. Loke. The socially networked car for safety, efficiency and the climate (abstract). In *Proceedings of the International Conference on Traffic and Transport Psychology (ICTTP)*, 2016.
10. Tim Roughgarden. *Twenty Lectures on Algorithmic Game Theory*. Cambridge University Press, 2016.
11. H. Seera, S. W. Loke, and T. Torabi. Towards device-blending: Model and challenges. In *Advanced Information Networking and Applications Workshops, 2007, AINAW '07. 21st International Conference on*, volume 2, pages 139–146, May 2007.

Chapter 4
Scalable Context-Awareness

4.1 Context-Aware Mobile Computing

Since the pioneering work on context-aware computing by Schilit et al. [12] over two decades ago, there have been tremendous developments in context-aware mobile computing [1, 7],[1] a mobile device is made aware of the current context of the user, including the circumstances or the surroundings as well as the user's activity on the phone, or the phone's current state (e.g., battery level, device properties and so on), and can take action based on such context information. Mobile sensing [4, 13] on the device is used to obtain information about the user, including the user's location, objects nearby (e.g., via WiFi or Bluetooth scanning) as well as the current physical activity of the user (e.g., walking, on a bus, etc), i.e. the work on mobile activity recognition (e.g., [14]), and the current user interaction with the apps on the phone (e.g., what the user is looking at). A large range of data analysis techniques has been employed to process sensor data in order to learn to recognise activities—recently, Deep Convolutional Neural Networks have been employed achieving accuracy in recognition of up to 97–99% [5].

Based on mobile sensing, there have been much work on reasoning with such mobile sensor data to infer higher-level user behaviours or activities.

For example, mobile intention recognition [6] aims to infer the user's intention from sensor measurements about the user's spatial behaviour (represented as trajectories of movements) and then help the user achieve the intention. The idea is to find the intention that best explains a given spatial behaviour of the user. For example, using a clothes shopping scenario, given sequences of recognised

---

[1]See videos on context-awareness at Google I/O 2015: https://www.youtube.com/watch?v=xgcj7VbDalk.

© The Author(s) 2017
S.W. Loke, *Crowd-Powered Mobile Computing and Smart Things*, SpringerBriefs in Computer Science, DOI 10.1007/978-3-319-54436-6_4

behaviours made up of searching, walking, sauntering, standing, picking an item or dropping an item, one can try to infer intentions such as *BuyPullovers*, *SearchShelf*, *TryClothes*, *LeaveFittingRoom*, *EnterFittingRoom*, *Queue* and so on. Spatially constrained grammars are used to describe spatial behaviour sequences.

The work in [2] attempted to infer high-level activities from low-level activities by processing mobile sensor data via a knowledge base, mapping high-level activities to low-level activities using Answer Set Programming. The idea is that an activity such as a *user-leaving-work* can be broken down to a set of activities including leaving the office, walking to the carpark, and getting into a car and driving off.

4.2   Larger Scale Sensing: Place Level Sensing

Once the data from crowds of people can be aggregated and processed and this data combined with sensor network data, a whole range of activities can be sensed accurately. The work in [11] employed sensor data from fixed ambient sensors in the smart home combined with sensor data from smartphones to recognise activities of daily living in the home.

The work in [9] provided a platform to recognising activities at a place by mapping processed sensor data to place activities described in an ontology. The platform can obtain sensor data from multiple devices and the data is then analysed to determine the activity happening at a place, and the time-stamped history of inferred activities at the place can be stored and queried. The idea of such work is to make places 'legible' for users, that they have a good picture of what is going on at the place.

4.3   Social Sensing

Social sensing [15] is defined 'as the act of collection of observations about the physical environment from humans or devices acting on their behalf', which includes the use of human contributions as well as mobile sensor data. An example are postings on Twitter on attacks in Syria in 2013 that provided a perspective on the situation as seen from the eyes of particular individuals. However, the idea of crowdsensing via automatic capturing of sensor data on mobile devices mentioned in Chap. 1 coupled with manually contributed social network postings can provide insightful situation-awareness, not only about what is observed but about the observers themselves. The reliability of the data obtained and how to aggregate such data become important issues.

The project Common Sense[2] aims to develop mobile sensing technologies to help whole communities obtain and analyse environmental data such as air quality. For air quality, mobile devices upload sensed data which are then aggregated to form a pollution map. There is a need for incentive mechanisms to compensate contributors, even if some are willing to do so for free, say for altruistic reasons. The idea is that ordinary citizens can participate using their own mobile sensors in monitoring and sensing applications that have benefits to the community itself. There is discussion on incentives for community sensing in [3].

The ability to combine resources from multiple 'sensors' means that information can be cross-checked and some degree of validation and verification of contributed information can be done, improving reliability and trustworthiness of contributed data. Peer prediction based verification or peer review might be used as a means to encourage or reward truthful reporting (e.g., [10], where rewards to workers are computed as a function of the worker's answer and other workers' answers, and consistency is rewarded, and also that answers which are rare yet consistent are rewarded even more), or to determine if crowdsourced contributions or crowdsensed data is valid or perhaps corrections or verification can be done by exploiting correlations among entities [8] or crowdsourcing the verification process itself.[3]

4.4 Scaling Up Context-Awareness

Sensing using mobile devices can start from individuals and then scale up to groups, aggregating the sensor data from a group of people, as mentioned in Chap. 2. The next step in scaling up is to sense crowds of people or people within a given community or at a place, e.g., crowdsensing as well as community and social sensing, and place level sensing. Many cities also have fixed sensors now, distributed throughout the city, including cameras. Indeed mobile sensing, mobile crowdsourcing, sensing using fixed sensors (and cameras), and social media knowledge extraction can be integrated and cross-analysed, in order to make sense of what is going on.

Consider a mobile app to enable one to find out or probe particular geographical areas. Figure 4.1 shows a map of an area, say within a mobile app, and suppose that the app enables a user to mark a circular region on the map (say centred at 3 km north from the user and radius 1 km) and then ask 'what is going on in that area?' or even a slightly more complex query like 'what is going on in this area relevant to me now?'. For example, the user marks three circular regions on the map as shown in the figure, one of which is around the user, and asks what is happening in those regions.

---

[2]http://www.communitysensing.org.

[3]http://blogs.oii.ox.ac.uk/policy/verification-of-crowd-sourced-information-is-this-crowd-wisdom-or-machine-wisdom/.

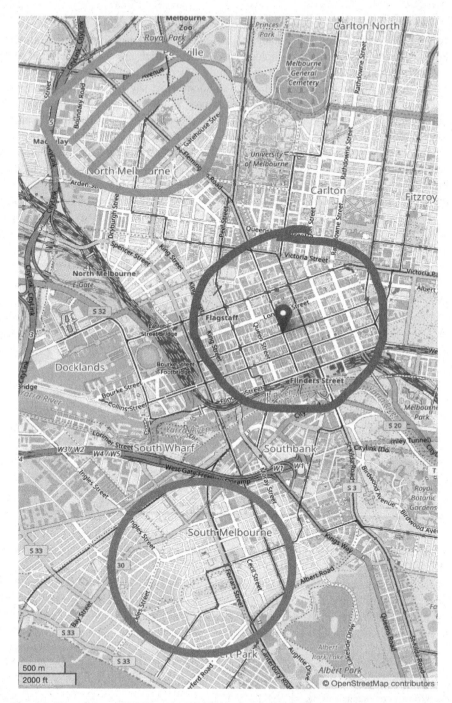

Fig. 4.1 Map of a part of Melbourne city with three regions of interest marked for querying. The region being queried could be one surrounding the user's location (as illustrated by the *circle in the middle*) or it could be a region centred at a point some distance from the user (as illustrated by the other two *circles*). (map src: http://www.openstreetmap.org)

The answer may be a wide range of information, including traffic conditions, events happening, where the crowds are and what they might be doing there, major incidents, and other commercial, tourist, education and social information. The user, ideally, should then be able to drill down to find very detailed information about an event of interest, and even connect and link to relevant parties from the app itself to find out even more or to make appropriate connections on demand. A combination of social media data, crowdsourced contributed information, as well as fixed sensor data might be employed for this purpose. Where demand is high and a drone could serve the information needs of a group of people, drones or ground robots might be employed to 'investigate'.

The user might also change the radius or the centre of the circle and in almost real-time, the relevant answer to the above questions is then updated for the new region on the map. Current technology can already be employed to implement such a mechanism. Augmented reality applications such as Layar[4] can already provide layers of information over the physical world that can be added or removed depending on the user's preferences. However, there is still work in exploring the range and level of detail of information that can be provided when combining myriad sources, and how best to correlate such information from multiple perspectives. Moreover, physical annotations within the region, managed via a crowd+cloud machine mentioned in Chap. 2, can be another source of information, in addition to Web-based social media data and sensor network data.

4.5 Scalable Context-Awareness for Smart Cars: A Use Case

The ability to understand the situation in particular regions can be used by smart cars in decision making, which we explore in this section, just as an example application.

A self-driving (or autonomous) smart car can be programmed with a particular destination, and it could bring the passengers there, but upon arrival, there are a number of possibilities, four of which are:

1. the autonomous car drops off the passengers (including the driver) and then either proceeds to find a car park nearby, or simply cruises around nearby;
2. the autonomous car tries to find a car park as near to the building as possible;
3. the autonomous car drops off the passengers and goes home (to come back later to pick up the passengers); or
4. the autonomous car allows the driver to take back control on nearing the destination.

Note that the last option would enable the driver to take over and it would proceed as a normal non-autonomous car with the driver making decisions about what to do from then onwards. The first two possibilities are less clear cut; if suppose the place

---

[4]http://www.layar.com.

turns out to have a huge traffic jam so that even simply dropping off the passengers would not be easy and the car might be stuck in traffic waiting for its turn at the drop-off zone. Also, if the car chooses to look for a car park, what would happen if it does not find a car park (or at least not a car park space near to the building the driver wants to go to)? The car would need to make some decisions on approaching the target building; it could wait to have its turn to drop off the passengers or it could go straight to find a car park and then passengers alike only after parking—the latter might take less time in case there is heavy traffic around the drop-off zone. A variant of the fourth option is to ask the passengers for further instructions about what to do. However, the car has to know when it needs to involve its passengers in such decision-making, if suppose the passengers are simply leaving it to the car to take them to the right place.

With scalable context-awareness, the car could probe the areas around the building at different levels of detail and granularity, a few minutes before arrival, in order to make its own informed decision about what to do or to inform the passengers of the situation (and ask them to help it make a decision about what to do). The car or passengers will need to know if parking further and walking would be more cost or time-efficient than say simply waiting in the car to be dropped off; situation-awareness of where the crowds are, the parking situation and the traffic situation are knowledge that could inform the car's decision.

Another possibility is for the smart car to take a more scenic route and arrive near the building a little later, knowing that the area around the building is currently congested anyway—which might lead to the best experience for its passengers (instead of arriving seemingly early at the area but having to wait long in the car to be dropped off). Of course, on the car's estimate of the wait time, the passengers might decide to alight and walk to the building. In perhaps a more contrived example, it could be that free parking is available after a certain time near the place the passengers want to go to, and the car could take a safer more scenic less congested route at the most economical (perhaps slower) speeds to get there so that fuel costs and parking costs might be saved, trading off a little time. A car might also decide to take the passengers through a route with smart advertisements delivered to users or with along-the-road e-signs in order to earn some 'eyeball revenue' for them, from advertising businesses. The third option could apply to save parking costs and reduce congestion if the passengers stay in the building long enough for the car to go home and come back (trading off fuel costs for the return trip).

We highlight the issue that a smart car could go beyond autonomous driving to actually making improved decisions for the user, or at least providing relevant suggestions (based on its awareness of the trip context or details of the travel situation at particular regions).

This scenario might involve sensors near the building or the exchange of situational information among cars for this to work. It remains a rather futuristic scenario if cars can make such decisions for users easily. There seems a complexity of travel decisions to be made, which detailed context-awareness could aid, beyond simply being able to self-drive from A to B.

Many other scenarios can be investigated in this setting, including logistics and delivery as well as optimising public transport and vehicle routing.

4.6 Summary

This chapter has outlined the notion of scalable context-awareness, where varying levels of details about physical world situations can be obtained, in order to provide better information for making decisions in everyday life, from travel to work. Details of how to build a wide area infrastructure for such environments would touch on smart city technologies in urban settings to socio-technical issues for end-users. Users of such scalable context-awareness apps will not just be persons with mobile devices but also perhaps smart vehicles aiding users.

References

1. Guanling Chen and David Kotz. A survey of context-aware mobile computing research. Technical report, Hanover, NH, USA, 2000.
2. Thang M. Do, Seng W. Loke, and Fei Liu. *HealthyLife: An Activity Recognition System with Smartphone Using Logic-Based Stream Reasoning*, pages 188–199. Springer Berlin Heidelberg, Berlin, Heidelberg, 2013.
3. B. Faltings, J. J. Li, and R. Jurca. Incentive mechanisms for community sensing. *IEEE Transactions on Computers*, 63(1):115–128, Jan 2014.
4. Geri Gay. *Context-aware mobile computing : affordances of space, social awareness, and social influence / Geri Gay*. Morgan and Claypool Publishers [San Rafael, Calif.], 2009.
5. Wenchao Jiang and Zhaozheng Yin. Human activity recognition using wearable sensors by deep convolutional neural networks. In *Proceedings of the 23rd ACM International Conference on Multimedia*, MM '15, pages 1307–1310, New York, NY, USA, 2015. ACM.
6. Peter Kiefer. Mobile intention recognition. In Steffen Hölldobler, Abraham Bernstein, Klaus-Peter Löhr, Paul Molitor, Gustaf Neumann, Rüdiger Reischuk, Myra Spiliopoulou, Harald Störrle, and Dorothea Wagner, editors, *Ausgezeichnete Informatikdissertationen 2011*, Lecture Notes in Informatics. Gesellschaft für Informatik, 2012.
7. Seng W. Loke. *Context-Aware Pervasive Systems*. Auerbach Publications, Boston, MA, USA, 2006.
8. Chuishi Meng, Wenjun Jiang, Yaliang Li, Jing Gao, Lu Su, Hu Ding, and Yun Cheng. Truth discovery on crowd sensing of correlated entities. In *Proceedings of the 13th ACM Conference on Embedded Networked Sensor Systems*, SenSys '15, pages 169–182, New York, NY, USA, 2015. ACM.
9. Tuan Nguyen, Seng W. Loke, Torab Torabi, and Hongen Lu. On the practicalities of place-based virtual communities: Ontology-based querying, application architecture, and performance. *Expert Syst. Appl.*, 41(6):2859–2873, 2014.
10. Goran Radanovic, Boi Faltings, and Radu Jurca. Incentives for effort in crowdsourcing using the peer truth serum. *ACM Trans. Intell. Syst. Technol.*, 7(4):48:1–48:28, March 2016.
11. Nirmalya Roy, Archan Misra, and Diane J. Cook. Ambient and smartphone sensor assisted ADL recognition in multi-inhabitant smart environments. *J. Ambient Intelligence and Humanized Computing*, 7(1):1–19, 2016.

12. Bill Schilit, Norman Adams, and Roy Want. Context-aware computing applications. In *IEEE Workshop on Mobile Computing Systems and Applications*, Santa Cruz, CA, US, 1994.
13. Sougata Sen, Archan Misra, Rajesh Balan, and Lipyeow Lim. The Case for Cloud-Enabled Mobile Sensing Services. In *Proceedings of the first edition of the MCC workshop on Mobile cloud computing*, MCC '12, pages 53–58, New York, NY, USA, 2012. ACM.
14. Muhammad Shoaib, Stephan Bosch, Ozlem Durmaz Incel, Hans Scholten, and Paul J.M. Havinga. A survey of online activity recognition using mobile phones. *Sensors*, 15(1):2059, 2015.
15. Dong Wang, Tarek Abdelzaher, and Lance Kaplan. *Social Sensing: Building Reliable Systems on Unreliable Data*. Morgan Kaufmann Publishers Inc., San Francisco, CA, USA, 1st edition, 2015.

Chapter 5
Drone Services for Mobile Crowds

5.1 The Rise of Drones

There have been tremendous recent developments in drone (unmanned aerial vehicles, or UAVs) technology [3] in terms of control and automatic flight of drones,[1] so that while regulations and drone protection measures (protecting drones and protecting people from drones) are being investigated,[2] practical applications are being developed, from product delivery, disaster assessment and recovery, art and exergaming [7] to self-driving cars guided by drones.[3]

This chapter discusses the idea of drone services, where drones are employed to deliver a range of services for users, either using drones owned by users themselves, or owned by a company or organisation.

5.2 Can We Imagine Drone Services?

While it is hard to conceive of billions of drones in the sky, and perhaps it is hard to imagine there being as many drones as there are mobile phones, companies using drones to provide services might become more commonplace if regulations and appropriate safety measures can be suitably created. More mini-drones or higher

---

[1]See for example, the work at http://www.kumarrobotics.org and at http://www.idsc.ethz.ch/research-dandrea.html.

[2]http://www.theatlantic.com/magazine/archive/2015/11/playing-defense-against-the-drones/407851/.

[3]http://phys.org/news/2016-12-ford-drones-self-driving-cars.html.

© The Author(s) 2017   47
S.W. Loke, *Crowd-Powered Mobile Computing and Smart Things*, SpringerBriefs
in Computer Science, DOI 10.1007/978-3-319-54436-6_5

attitude drones might accommodate more drones in the sky but highways in the sky[4] and virtual paths [5] might be needed to impose order and structure in the sky (for safety, aesthetics, efficiency and so on).

A company might provide a phone-a-drone service for people out surfing or people doing a bush walk who want photographs taken from 'impossible' angles (e.g., from multiple angles concurrently while in the ocean or from off the edge of a cliff). A crowd of cars stuck in a traffic jam might decide to cooperate (and co-hire) to send a drone to investigate and then share the information.

Consider the (fictitious) scenario illustrated in Fig. 5.1, with three drone stations D1, D2 and D3 where drones can be launched (and where they are charged and managed) to serve user (or client) requests. User requests are illustrated in the figure as red spots. Suppose there are k drones at those stations, then given the requests at any one time, where each request has a location and a time for fulfilment, an algorithm is needed to schedule the drones to meet the requests on time, or at least to meet as many requests as possible within the time constraints, given that drones have limited power and needs to be recharged as well.

Drones come in a range of sizes and form factors. Hence, the applications of drones can be as diverse as the types of drones. There will be general purpose hobbyist drones but there can be (and are already) custom-built, adapted or specialised drones for particular services and functions. A swarm of drones for a lights display[5] might be built differently from drones forming the building blocks of a 3D flying display.[6] Drones used for movie filming[7] might be different from those that can perch on street lights and repair them.[8] Hence, while we can talk about drones for rental for general hobbyist applications, there could be large vertical segmentation of application-specific drones specialised to deliver particular services.

---

[4]http://www.wbur.org/onpoint/2016/01/04/nasa-drone-highways.

[5]At http://www.livescience.com/56792-drone-fireworks-show-world-record.html is mentioned a 500 drone fleet light display.

[6]See BitDrones at http://interactions.acm.org/archive/view/may-june-2016/bitdrones.

[7]https://skytango.com/drones-in-movies-7-hollywood-movies-filmed-with-drones/.

[8]This is a funded project at the University of Leeds: http://www.popularmechanics.com/technology/robots/a18051/leeds-is-trying-to-use-drones-to-become-the-first-self-repairing-city/, though not without its critics: http://spectrum.ieee.org/automaton/robotics/industrial-robots/you-probably-shouldnt-expect-city-repairing-drones-any-time-soon; it is also reported that Amazon has been granted a US patent to allow drones to dock at street lights—perhaps an ingenious way to combine drone stations and widely available infrastructure, at http://luxreview.com/article/2016/07/amazon-drones-to-dock-on-streetlights.

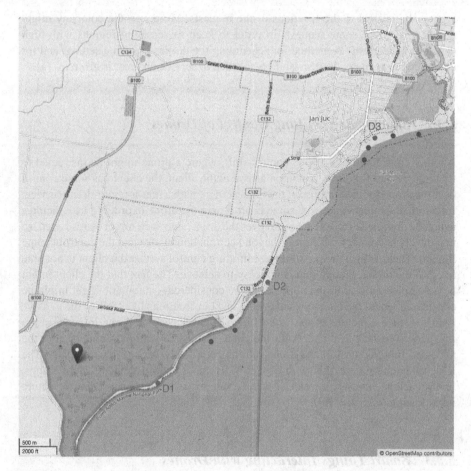

Fig. 5.1 Map of a part of Victoria near Torquay and Bells Beach, a spot for surfing, showing, fictitiously, locations of three drone stations, the spots labelled D1, D2 and D3 and the other spots denote requests (i.e., the points where drones have been requested by a client). (src: http://www. openstreetmap.org)

5.3 Issues and Challenges

5.3.1 Scheduling Drones

While the scenario in Fig. 5.1 is contrived, different applications could have the same fundamental structure. The idea is that the strategy where drones meet requests in a greedy manner, i.e., each drone addresses requests as they come and addresses the request nearest to it first, could result in drones ending up in one part of the region, and then needing to travel much further to service requests in another region. Instead, at the risk of being late, some requests might be serviced by drones which are not the nearest to the locations of the requests. Moreover, the company might decide to give up some requests in order to keep to servicing regions with high densities of requests. A method for scheduling the drones in such a setting, and for computing how many drones would be needed to service an area, is given in [4].

5.3.2 Sharing and Shifting Control of Drones

Another consideration is drone control shift, where a drone might be instructed to fly to the point of request, and upon arrival at the client, the client has some control over the drone for the duration of the request (e.g., with a rent-a-photo-drone service, for 5 min, the client is allowed some control over the drone in order to take pictures using the drone). The client's control over the drone, however, might be restricted, so that the client cannot fly the drone beyond certain boundaries and the control is time limited. There is also the possibility of limiting control so that the client is not able to crash the drone, though that is not easy to achieve. The fact that the client might be a complete novice drone pilot should be considered—automatic flight might be considered where the client provides high level instructions like 'follow me', 'move right' or 'take a picture of me when I do...' and leaves the drone to perform the instructions as best as it can. After the client is finished with the drone or after the allocated time, control of the drone returns completely back to the company. The issue of human-interaction with drones via remote control devices, mobile apps, brain-machine interfaces, voice, gestures and so on remain an avenue for future investigation (e.g., see [6]).

5.3.3 Smart Things Interacting with Drones

Will your fridge talk to drones? An interesting issue to consider is how drones will interact with other drones and with the Internet-of-Things mentioned earlier. We have already mentioned how cars might use and talk to drones in order to gain awareness about traffic situations. A similar idea can be applied to pedestrians or walkers who want to increase situation-awareness about a region 1 km from where they are or about the area around the corner of a large building, or for drones who want to have situation-awareness further ahead from where they are (i.e., drones get information from other drones ahead of them in a path).

But other interactions are possible. A fridge might order items that have ran out and the drone delivers the items informing the fridge that they have been delivered, and perhaps automatically dropping them into the house and the items are then channelled directly into the fridge via the fridge's backdoor—you don't even realise the items have been replenished until you see the bill. As fanciful as the scenario is, the idea that drones can interact with smart things creates new possibilities. There are, of course, security issues, as drones might be used to hack the Internet-of-

Things.[9] In the Internet-of-Things map project by Praetorian,[10] drones are used to detect where smart networked things are by scanning for devices using the ZigBee communication protocol. A potential usage of the device discovery and communication, however, is that drones can discover and be guided by ZigBee beacons, and interact with ZigBee devices on the ground obtaining new information or provide information to such devices, forming an ecosystem of collaborative devices.

5.3.4   Infrastructure to Support Drone Services

The infrastructure for massive deployment of drones is certainly another issue to be considered. Companies such as Amazon which aims to use drones for delivery would also need to (or likely already doing so) consider how such drones can be supported in this function—e.g., how homes or offices might be instrumented or what public infrastructure needs to be available for that to work. Such infrastructure, including drone stations, might be privately owned by home owners or businesses or some might be public infrastructure, in the category of roads and street lights, to be used by multiple (perhaps licensed) service providers. There are issues whether private drone operators would be able to use such shared infrastructure (e.g., an infrastructure of thousands of drone stations—perhaps deployed on street lights and elsewhere), or even whether such shared infrastructure would be shared (or rented out to other parties for a fee). Government might also find such infrastructure handy, for drones used in various public service operations, including emergency services by police and fire crews.[11] As experience with drone services increases, more general architectures, platforms, and techniques for their implementation might be developed, as well as greater specialisation of drone capabilities.

Other kind of 'infrastructure' might be less visible such as public highways in the sky for drones.[12] For drones to work safely beyond line-of-sight, LTE type cellular communication might be feasible.[13]

Drones might also be supported by an information infrastructure. As mentioned in Chap. 2, the idea of annotations 'attached' to (e.g., Bluetooth or ZigBee tagged) objects and places could yield a scenario of layers of 'semantic' information [1] that not only humans can retrieve on their mobile devices but drones can obtain, process and use. The idea is similar to augmented reality where people can see

---

[9]http://thehackernews.com/2015/08/hacking-internet-of-things-drone.html.

[10]https://www.praetorian.com/iotmap/#14/30.2679/-97.7440.

[11]http://news.sky.com/story/drones-saving-lives-of-emergency-workers-10296876.

[12]https://www.qualcomm.com/news/onq/2016/11/09/path-5g-building-highway-sky-autonomous-drones.

[13]https://www.qualcomm.com/invention/technologies/lte/advanced-pro/cellular-drone-communication.

Fig. 5.2 Illustration of physical annotations associated with places in the real world

virtual information or graphics superimposed on top of real-time images of the real-world as seen via a camera, so that if a person points his/her camera phone on a building, s/he might see information about the building superimposed on the camera image. The physical annotations can be associated with places via GPS coordinates (geotagging the annotation with the GPS coordinates of the place and storing this in a database) or via other electronic tagging mechanisms. In this case, a drone flying over the building can retrieve information about the building and use it to guide its operations. Figure 5.2 illustrates the idea of an annotated world with annotations that can be read by humans but also potentially usable by drones to help their operations. For example, one could command the drone to 'drop this package at the home where I grew up' giving a rough address but the drone on arrival finds the particular house with that annotation and then locates the destination. Such annotations might also be useful at night when drones have limited night-vision and do not have camera-view visibility, and in general, for drones to make better sense of the physical world.

5.3.5   Drones from the Crowd

What about drone sharing? We discussed the idea of thing sharing, where vehicles to umbrellas might enter the sharing economy. If people and businesses have their drones, will multiple businesses or a crowd of people be able to pool their drones together to form a fleet of drones that can be shared and used opportunistically? Can there be a Uber-like drone sharing economy, e.g., I allow my drone to be used for delivering packages when I am not using it? Can drones be repurposed temporarily, for example, in an emergency situation, to help others? There remains a question of 'community drones' and a 'we'-centric computing model mentioned earlier, where one buys a drone but can be repurposed for community use when needed.

5.4 Summary

While we proposed and outlined the potential of drone services, and there have been tremendous developments in the versatility and in the robotics engineering concerning drones, there are a range of challenges. There are a number of inter-related issues to consider, including:

- drone numbers and their management for adequately providing a service
- legal framework and regulations for (all and specific) drone services (e.g., emergency drones should be clearly marked as such, to be separate from private drones, etc.)
- drone safety and security, client safety, general safety (including impact to third parties of drones delivering services to clients); it is not just safety but aesthetics, noise and other risks that must be considered even if people are generally protected from bodily harm from drones
- restricted drone control and shift of control of drones, including drones with automatic flight but high-level commands, and comfortable human–drone interaction (especially for non-technical users to use certain drone services where they may need to interact with drones, e.g., photo-taking)
- serviceability of requests
- scheduling of drone flights and time allocation to requests
- positioning of drone stations
- the range of services a company's drones provides
- mechanisms for effective drone sharing
- how drones can collaborate with each other and with the Internet-of-Things to service requests efficiently
- suitable (public and private) infrastructure to support multiple drone applications

Other issues of maintaining information relay networks when drones are moving, robustness, trust, and handling drone failures are also noted in [2], particularly in the case of using drones for disaster management. For drones, it seems that the sky is the limit, the range of applications being investigated has not stopped growing, and the excitement about drones in recent years has continued to grow. Engineering, algorithmic, legal, and socio-technical challenges remain for drones to be a robust technology and for drone service provisioning to be a viable business model.

References

1. Ahmad A. Alzahrani, Seng W. Loke, and Hongen Lu. An advanced location-aware physical annotation system: From models to implementation. *J. Ambient Intell. Smart Environ.*, 6(1): 71–91, January 2014.
2. Milan Erdelj, Enrico Natalizio, Kaushik R. Chowdhury, Ian F. Akyildiz, undefined, undefined, undefined, and undefined. Help from the sky: Leveraging uavs for disaster management. *IEEE Pervasive Computing*, 16(1):24–32, 2017.
3. Seng W. Loke. The internet of flying-things: Opportunities and challenges with airborne fog computing and mobile cloud in the clouds. *CoRR*, abs/1507.04492, 2015.
4. Seng W. Loke. Smart environments as places serviced by k-drone systems. *JAISE*, 8(5): 551–563, 2016.
5. Seng W. Loke, Majed Alwateer, and Venura S.A. Abeysinghe Achchige Don. Virtual space boxes and drone-as-reference-station localisation for drone services: An approach based on signal strengths. In *Proceedings of the 2Nd Workshop on Micro Aerial Vehicle Networks, Systems, and Applications for Civilian Use*, DroNet '16, pages 45–48, New York, NY, USA, 2016. ACM.
6. Ekaterina Peshkova, Martin Hitz, and Bonifaz Kaufmann. Natural interaction techniques for an unmanned aerial vehicle system. *IEEE Pervasive Computing*, 16(1):34–42, 2017.
7. Jurgen Scheible, Markus Funk, Klen Copic Pucihar, Matjaz Kljun, Mark Lochrie, Paul Egglestone, and Peter Skrlj. Using drones for art and exergaming. *IEEE Pervasive Computing*, 16(1):48–56, 2017.

Chapter 6
Social Links for Crowds and Things

This chapter considers social links formed within crowds of people, often captured digitally in social media networks, including the notions of 'following', 'being followed', 'friends', and 'connections'. The kind of links within crowds are diverse. This chapter examines three ideas, the notion of favour networks within crowds, automatic social networking, and social networks for things.

6.1 Favour Networks in Mobile Crowds

Mobile device clouds or crowd computing involves mobile devices providing services to one another or doing a favour for another (perhaps altruistically or for something in return, even a favour in return), where a mobile device can be smartphone, smartwatch, a drone or even a smart car. The idea is that not only can a mobile device utilise cloud resources but they can act as resource providers, or more generally, fulfil favours for one another. Resources they provide or share might be:

- machine computation based, that can range from idle CPU cycles, temporary storage, access to GPS, Internet connectivity (in the tethering style), sensing capabilities, or even screen estate [1],
- human computation based, where the user of the mobile device performs a tasks manually (e.g., as in much work on mobile and spatial crowdsourcing), and
- mixed human and machine computation based (e.g., a car with the approval of its owner may allow another car to park in a given car slot).

In particular, with developments in device technology and in short-range networking technologies such as Bluetooth, WiFi-Direct, LTE Direct and so on, a device could be quite well-served by devices nearby—the notion of a cloud of

© The Author(s) 2017   55
S.W. Loke, *Crowd-Powered Mobile Computing and Smart Things*, SpringerBriefs
in Computer Science, DOI 10.1007/978-3-319-54436-6_6

nearby mobile devices providing resources has been explored in much work. A lot of such work on spatial or mobile crowdsourcing involves payment incentives and involves a centralised platform for task assignment to suitable workers.

We consider in this section the scenario of peer-to-peer favour networks, where a centralised platform might be involved but mainly for accounting or validation purposes, while most of the incentives for sharing is based on a favour giving and receiving system among nodes that encounter each other (and interact with each other mainly via short range networking technologies such as Bluetooth or WiFi-Direct, and perhaps LTE Direct in the future). Tit-for-tat or quid pro quo style systems can be useful for peer-to-peer sharing of resources involving reciprocation of favours. However, sometimes a device utilises the resources of another device but may not meet the same device till much later or even never. However, it could be possible that a network of favours form, e.g., device i uses resources of j and j uses that of k which then uses that of i, in which case every device does a favour for another but is also done a favour by another. This could work even when no direct reciprocation happens (e.g., j never uses any resource of i). Hence, a favour doing system where direct reciprocation is not necessary might be more generally useful and feasible.

The notion of a sustained network of peers in a favour network has been explored in other contexts, e.g., [4], where there are built-in penalties for not performing a favour, so that a node which refuses a favour has its links to its neighbours severed (effectively the node is ostracised) costing it to forego potential future benefits of having links. But if a node refuses a favour and has links to its neighbours severed, the neighbours now have fewer links and so tend to lose less if they, in turn, refuse favours. In some cases, this could lead to a collapse of the network since many nodes then have less to lose if they have few links in the network and then start to refuse favours. It was shown that a network of the form of a *social quilt* is resistant to collapse and robust when nodes start to refuse favours. Such a social quilt is formed by a combination of m-cliques where each m-clique is a complete network of $m + 1$ nodes where each node has exactly m links. The idea of having m links for a node discourages the node from refusing favours since it has the benefits from m links to lose if it does so and the idea of cliques is that damage is localised (if links were to be severed, they typically occur within a clique).

Consider a situation where refusing a favour or behaving badly towards another might potentially weaken the link or adversely affect a relationship. For example, consider a simple scenario illustrated in Fig. 6.1 where A asked B to do a favour for C and B agrees to do so not wanting to antagonise A or lose the link with A.

If the situation can be quantified, then suppose the cost to B of doing the favour is c and the value of its link to A is L, then provided $L > c$, the favour is worth doing. Another perspective is that, there is an accountability of B to A about actions that B makes to C. For example, B might behave badly towards C obtaining a benefit b, but A might come to know about it, and so, B loses L, and if $L > b$, then B should think twice about its action towards C. The link itself might not be severed but perhaps the value diminished—e.g., by a fraction δ, so that $\delta L > b$ guarantees B's good behaviour towards C if B is rational.

Fig. 6.1 Illustration of three people where A is friends with both B and C and A asked B to do a favour for C. Even though B did not know C previously, B might oblige since B does not want to lose or weaken the link with A

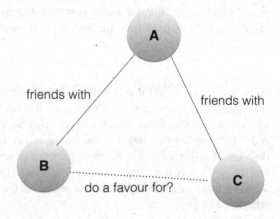

Also, later B might not mind asking A to help another friend D, in return. In general, in a community where people know each other and there is potentially long lasting links, we would expect such reciprocation as well as effort put in for the maintenance of valuable links.

We consider a network where the nodes are mobile devices (and users) and a link between two nodes indicates that one node can *potentially use* the resources of another. The links are also directed so that $i \rightarrow j$ means i *can do a favour for* j, and not vice versa, and both $i \rightarrow j$ and $j \rightarrow i$ means i can do a favour for j and j can do a favour for i (denoted by $i \leftrightarrow j$). And the idea of j can do a favour for i could mean, for example, that j allows i to use its resources (which would normally be within certain limitations but we simplify by ignoring details of resource usage here), whenever i and j come near enough to each other (e.g., within WiFi-Direct range).

This means that initially, we start with a complete network of $\leftrightarrow$ links, assuming that any node can serve or be served by any other node potentially, and that initially every node is participating in the favour giving system—we will then explore what happens when nodes begin to refuse favours when asked (i.e., when they stop participating). Similar to [4], we assume that there is a cost of c of providing a favour to another node when asked and a node receives a benefit of b from being served by another node. So, in a period, suppose one node asked a favour of another and is granted, then the receiving node gets a benefit of b and the giving node incurs a cost c. But the cost could depend on the type of favour being asked, we assume for simplicity that the same cost is incurred regardless of the type of favour (or consider here only one type of favour). Also, we assume that a node i wants a favour from another node j with probability p (and a node j wants a favour from i with probability p), i.e., a link $i \leftrightarrow j$ conveys to both its ends an expected utility of $p \cdot (b - c)$ in each period, assuming one favour could be asked by a node in each period and one favour could be asked of a node in each period.

With discounting using discount factor $0 < \delta < 1$, this benefit is less in the next period $\delta \cdot p \cdot (b - c)$, and in the period after the next, we have $\delta^2 \cdot p \cdot (b - c)$, and so on, which means that for a society with only two nodes, the discounted stream of utility expected by each from the link between them is given by:

$$(1 + \delta + \delta^2 + \ldots) \cdot p \cdot (b - c) = \frac{p \cdot (b - c)}{1 - \delta}$$

Now, suppose a node i not granting a favour when asked causes i to be ostracised (meaning it no longer can be served) and it refuses to serve others from here on, i.e., all its links to other nodes severed, if a node provides a favour at cost c in order to maintain this link for the future utility it can bring, it must be that the discounted utility of this link, starting from the next period, is more than the cost, i.e., $\delta \cdot \frac{p \cdot (b-c)}{1-\delta} > c$. But if the node has $d \geq 1$ links to maintain in a society with more than two nodes, then doing the favour preserves the total utility from d links, that is, when $d \cdot \delta \cdot \frac{p \cdot (b-c)}{1-\delta} > c$.

Typically, each node typically has a mobility pattern [2, 7], reflecting a node's daily routine and regular movement behaviour, and so, each node tends to come near to certain other nodes more often than some others. This means that in a network of n nodes, a node which refuses a favour asked of it and has all its links severed does not lose benefit equivalent to benefits from $(n - 1)$ links since, in practice, a node i might only encounter a fraction $0 \leq \epsilon_i \leq 1$ of other nodes in its lifetime (assuming that favours can only be given when a node encounters or is near enough to another node). A highly mobile node in a high density area might encounter a larger fraction of nodes. That is, if a node i does not grant a favour, suppose it loses all its links, that is, it loses the following expected utility $\epsilon_i \cdot (n - 1) \cdot \delta \cdot \frac{p \cdot (b-c)}{1-\delta}$, and not $(n - 1) \cdot \delta \cdot \frac{p \cdot (b-c)}{1-\delta}$.

It might be that the benefits and costs depend on the node itself, so that the benefit of a favour to a node i is b_i and the cost is c_i, that is, for each node i, we have $\epsilon_i \cdot (n - 1) \cdot \delta \cdot \frac{p \cdot (b_i - c_i)}{1-\delta} > c_i$ to make it worth doing the favour at cost c_i. (Note we will drop the subscripts in discussing a generic agent.)

If a node i does not grant a favour, suppose the penalty is that it loses all its links forever (similar to the 'grim trigger' notion), that is, it loses the following expected utility G_{grim}:

$$G_{grim} = \epsilon_i \cdot (n - 1) \cdot \delta \cdot \frac{p_i \cdot (b_i - c_i)}{1 - \delta}$$

where n denotes the total number of nodes participating in the network.

Now suppose that a node that sits out (i.e., does not grant a favour and is ostracised) not forever but for a *penalty duration* τ. The discounted stream of utility expected from a link over duration τ is:

$$(1 + \delta + \delta^2 + \ldots + \delta^\tau) \cdot p \cdot (b - c) = (1 - \delta^{\tau+1}) \cdot \frac{p \cdot (b - c)}{1 - \delta}$$

and so, not granting a favour causes i to lose utility:

$$G_\tau = \epsilon_i \cdot (n - 1) \cdot \delta \cdot (1 - \delta^{\tau+1}) \cdot \frac{p \cdot (b - c)}{1 - \delta}$$

Table 6.1 Payoffs for
Groups A and B when they
participate and when they sit
out

		Group A	
		Participate	Sit out
Group B	Participate	4,4	2,3
	Sit out	3,2	3,3

Note that $G_\tau < G_{grim}$ which means that limiting to a finite period of ostracism is less a deterrent than the grim trigger technique since grim trigger has more to lose.

Hence, a functioning (i.e., where nodes do not refuse favours) favour network can form even when nodes are mobile as long as they encounter enough of other nodes so that it is worth being in the network (and provided a node does want favours from time to time). Clearly, if p_i is higher for a node, and if n is larger, and ϵ_i is high, then there is more to lose with the grim trigger type penalty.

In the same way, given that there are already lots of nodes participating in the favour network so that n is high, then there is more incentive for someone to join. But a critical mass of participating nodes might be required before the incentive is big enough for a node to join. This is similar to the stag hunt game in Game Theory (Table 6.1). Consider two groups of nodes, A and B. If both participate, value is created and the payoff is 4 for both groups, but if both sit out, both have a smaller payoff of 3, but if only one group participates, and the other sits out, then the group that sat out maintains a payoff of 3, but the group which participated does not find enough value in participating but still incurred overheads in participating (since it is too small compared), and has a payoff of 2. Other incentives might be required to trigger participation initially until a critical mass is formed. Also, given a penalty duration τ which is not forever, some agents might only participate for a period leading up to the time they need a favour.

Another point to note is that the penalty of ostracising nodes can affect the entire network, i.e., when a node is ostracised, its potential benefit to others is lost, and the value of the network drops a little for all, decreasing the robustness. As mentioned earlier, a social quilt structure comprising a network of cliques is then useful here, so that a node that is ostracised might mainly diminish the benefits to its local clique, and the rest of the network remains robust.

6.2 Automatic Social Networking

From a brief overview of issues in favour networks in crowds above, we consider the problem of link formation. Most social media systems such as Facebook and LinkedIn are based on manual linking initiated by individuals, but they also have a 'people you may know' feature, or link recommendations. For example, from

Facebook,[1] 'People You May Know are people on Facebook that you might know. We show you people based on mutual friends, work and education information, networks you're part of, contacts you've imported and many other factors'. Facebook seems to have used collocation as a means for suggesting links but no longer (e.g., since people who have been collocated or who have spoken to each other might want to remain anonymous to each other).[2]

In [3], three main hypotheses about link formation are noted:

- *social foci*: links are more likely to form among individuals who share a social focus, e.g., co-workers, classmates, gym-mates, church congregation, and so on;
- *triadic closure*: links are more likely to form among individuals who have a common friend or know someone in common; and
- *homophily*: links are more likely to form among individuals who share the same social characteristics or are similar, such as physical appearance, age group, values, and cultural background.

Figure 6.2 shows a LinkedIn graph using the author's connections, created using a tool called Socilab,[3] where the nodes of the same colour are for people in the same country. Collocation is indeed a good heuristic for predicting links though not without its privacy issues. Also to note is that many of the connections are themselves connected to each other.

Link suggestion or prediction algorithms could and do use the above ideas to increase the quality of suggestions. After an extensive survey on link recommendation, the work in [5] proposes that utility of the potential link, and diversity should be considered for link recommendation. A link is recommended if it is predicted that it would have value to the user—how such value can be estimated is in general non-trivial. Diversity in the links can be measured in various ways, e.g., using the properties of individuals currently linked to and the individuals being considered for linking and a link might be recommended if it is to someone quite differently from those currently linked to. However, the value of such diversity might be application-specific. The cost of forming and not forming the links can also be taken into account, e.g., as in the earlier discussion, where the cost of not linking could mean not being able to exploit certain benefits.

Going beyond collocation, automatic linking could mean that a wearable system that monitors who we talk to physically and who we meet could then find out about the persons and then attempt to form a digital link with them in some social media platform. As mentioned, there could be reasons why people do not want such links

---

[1]https://www.facebook.com/help/www/501283333222485.

[2]http://fusion.net/story/319712/facebook-now-says-phone-location-not-used-to-recommend-friends/,   https://www.theguardian.com/technology/2016/jun/29/how-does-facebook-suggest-potential-friends-not-location-data-not-now; for LinkedIn link suggestions, see https://www.linkedin.com/help/linkedin/answer/29/people-you-may-know-feature-overview?lang=en,   and other articles on it, e.g., http://www.zdnet.com/article/how-linkedins-people-we-may-know-feature-is-so-accurate/.

[3]http://www.socilab.com/.

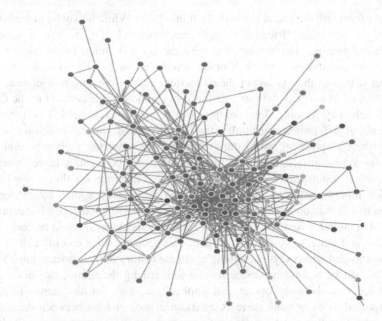

Fig. 6.2 A graph showing the author's LinkedIn connections showing that a majority of the connections are from the same country, and many are connected to each other

to form, but there are contexts where this might be useful, e.g., in a conference (to replace the exchange of business cards). Context could drive the utility and relevance of links, and some temporal and location constraints as well as user preferences could determine when such a system should be active.

6.3 Social Networks for Things

So far, our discussion mainly focused on people or people with their mobile devices. However, there could be social networks for things, quite separate from the social network for people. The social IoT concept[4] proposes that things could form a social network on their own, though governed by rules from humans. The social IoT does not require that things have agency and be able to act within their environment, but things with agent capabilities such as autonomy, rationality, communicative abilities, and proactive behaviour might be a welcome development, and such things will interact with other things, so that digitally representing and recording such links in their own social network of things can be useful.

---

[4]http://www.social-iot.org.

What form will the social network for things take? While it may be metaphorical to apply notions like 'friendship' and 'relationship' to things, it may serve an operational purpose, and provide intuitive ideas for how things might interact with people and collaborate with each other. Things might be able to find new links to other things as they go about their function, and by creating new connections, indirectly, build new capabilities. Also, things should be able connect to the Cloud and to each other forming clouds of things. A vision for the social IoT is articulated in the social IoT project,[5] including the idea of virtual objects, which are virtual counterparts of physical world things and devices, allowing their reasoning and functions to be digitised, and for links among things to be digitally represented.

There are a number of scenarios where interactions among things would call for their representation as links and it would be useful for such interactions to happen within the context of predefined links (or social relationships). For example, things that interact often with each other (and in helpful ways) might be considered 'friends', and when represented as such carries semantics that can influence future interactions and which says something about the history of their interactions. While 'friends' can be a good metaphor for the link among the things, this description of the link can also have operational implications, e.g., that the communications are expected to be helpful, there is a certain degree of trust between them (e.g., due to the fact that previous interactions were not malicious and were helpful) and that certain functionalities are expected between them (e.g., they tend to share resources whenever possible). Likely, other kinds of relationships among things can be represented. Ideas from ecology might be useful, where such relationships might be characterised as symbiotic or co-evolutionary (but hopefully not parasitic as such). Other types of relationships among things such as 'substitute', 'enhance', 'complement', 'enable', 'disable' and so on are discussed in [6], but there is likely a need for more carefully defined relationships that may be application-specific.

Links with well-defined semantics can expedite interaction among things since particular assumptions or expectations can be taken as given—for example, if the link is a 'friend' link, previously exchanged information might apply in the current interaction. Reciprocative relationships among things can also be captured within the notion of 'friends'—favours from one side to the other are recorded in the links themselves, and when the interactions among two things are not as expected, we might consider the notion of how links among the two things might become 'strained' (e.g., one of the two things has been hacked and is now behaving strangely which the other detects and then eventually severes the link). There could be applications to security where certain types of well-defined links among things imply that only certain interactions or communications are expected, providing a filter for strange or unexpected behaviours.

For cars, one could envision social networking among cars, as mentioned in Chap. 3. Given people's regular movement behaviours, we expect some cars to 'encounter' each other or be within kilometres of each other quite often.

---

[5]See the tutorial at http://www.social-iot.org/index.php?p=pubblicazioni&pu=14.

For example, my car and my neighbour's car actually 'live' nearby and we may travel to the same places during weekends (e.g., a nearby shopping centre) and even along similar routes (if we stay in the same suburb and work in the same area, such as the Central Business District). This is a variation of the 'familiar strangers' phenomenon, but for cars. The cars might find it useful to link with each other, to share route information or safety information, whether I know my neighbour personally or not. Social networks might also form among things such as appliances, and other devices, facilitating the kind of cooperation described in Chap. 3. Also, social links among things might be used as a heuristic to suggest social links among the things' owners.

6.4 Summary

For crowds of people and crowds of things, digital representations of their relationships in social links not only can mirror their interactions in the physical world but also can enhance their interactions. Such social networks not only capture the nature of the connections between things and people, but can provide a context for how they interact or how they should interact.

References

1. Frank H.P. Fitzek and Marcos D. Katz. *Mobile Clouds: Exploiting Distributed Resources in Wireless, Mobile and Social Networks.* Wiley Publishing, 1st edition, 2014.
2. Marta C. Gonzalez, Cesar A. Hidalgo, and Albert-Laszlo Barabasi. Understanding individual human mobility patterns. *Nature*, 453(7196):779–782, June 2008.
3. César A. Hidalgo. Disconnected, fragmented, or united? a trans-disciplinary review of network science. *Applied Network Science*, 1(1):6, 2016.
4. Matthew O. Jackson, Tomas Rodriguez-Barraquer, and Xu Tan. Social capital and social quilts: Network patterns of favor exchange. *American Economic Review*, 102(5):1857–97, May 2012.
5. Zhepeng (Lionel) Li, Xiao Fang, and Olivia R. Liu Sheng. A survey of link recommendation for social networks: Methods, theoretical foundations, and future research directions. *CoRR*, abs/1511.01868, 2015.
6. Harinder Seera, Seng Wai Loke, and Torab Torabi. Towards device-blending: Model and challenges. In *21st International Conference on Advanced Information Networking and Applications (AINA 2007), Workshops Proceedings, Volume 2, May 21-23, 2007, Niagara Falls, Canada*, pages 139–146, 2007.
7. Lijun Sun, Kay W. Axhausen, Der-Horng Lee, and Xianfeng Huang. Understanding metropolitan patterns of daily encounters. *Proceedings of the National Academy of Sciences*, 110(34):13774–13779, 2013.

Chapter 7
Conclusion and Future Work

Against the backdrop of technology trends such as cloud computing, IoT, mobile and wearable computing, crowd computing, a culture of sharing, collective computing, and swarm dynamics, this book has attempted to focus on five ideas, namely, crowd+cloud machines, extreme cooperation with smart things, scalable context-awareness, drone services for mobile crowds and social links in (mobile) crowds. We have not been exhaustive in our exploration of these areas, but we have sought to draw on current literature as well as the author's own work, to highlight ideas not yet adequately explored and to identify synergies, links, issues and challenges.

These five ideas are interlinked, though it has been convenient to discuss them separately. While they relate mainly to mobile computing, they also relate to people and devices in the non-mobile context. Social links may form taking into account the contexts of users and crowd+cloud machines developed to serve users might employ drones, which cooperate with hundreds of other smart things to work. Smart things themselves might *crowdsource* (to humans and other smart things) tasks that they find difficult, or *cloudsource* to find more resources to augment their capabilities.

The label *crowd-powered mobile computing and smart things* may conjure up a picture of crowds of people with devices forming virtual crowd+cloud machines that cooperate on a large scale over space and time, with social links digitally formed and represented, interconnecting agents (people and things), providing value to the crowds or the things themselves. This book aimed to outline what that picture might look like in terms of current technological trends, but the future of interconnected crowds is expected to be one with a much higher density of people and things, with enormous complexity of links, and so, providing enormous possibilities.

© The Author(s) 2017
S.W. Loke, *Crowd-Powered Mobile Computing and Smart Things*, SpringerBriefs
in Computer Science, DOI 10.1007/978-3-319-54436-6_7

Printed in the United States
By Bookmasters